QUESTIONING NATURE

QUESTIONING NATURE

British Women's Scientific Writing and Literary Originality, 1750–1830

Melissa Bailes

UNIVERSITY OF VIRGINIA PRESS
Charlottesville and London

University of Virginia Press
© 2017 by the Rector and Visitors of the University of Virginia
All rights reserved
Printed in the United States of America on acid-free paper

First published 2017

ISBN 978-0-8139-3976-6 (cloth)
ISBN 978-0-8139-3977-3 (e-book)

1 3 5 7 9 8 6 4 2

Library of Congress Cataloging-in-Publication Data
is available from the Library of Congress

Cover art: Portrait of Maria Banks Woodley Riddell (1806), by Sir Thomas Lawrence. (©National Trust/Angelo Hornak)

CONTENTS

✛

ACKNOWLEDGMENTS

Several fellowships provided time and resources that aided in this book's completion. I am especially grateful to the Huntington Library in San Marino, California, as well as Tulane University and the Dean of the School of Liberal Arts, the American Association of University Women, the University of Illinois at Urbana-Champaign, and the Chawton House Library. Early versions of chapters 2, 3, and 6 appeared in *European Romantic Review* 20.2 (2009), *Eighteenth-Century Life* 33.3 (2009), and *ELH: English Literary History* 82.2 (2015), respectively. While writing *Questioning Nature,* I benefited immeasurably from the conversation and advice of colleagues. I particularly wish to thank Ted Underwood, Justine Murison, Jennifer Keith, and Molly Rothenberg for their wonderful insights, support, and guidance over the years. Robert Markley, Gillen Wood, Ed White, Thomas Albrecht, and Gaurav Desai all read portions of this project and provided very helpful feedback. For their steadfast friendship, humor, interest, and encouragement, I'm grateful to Allison Davis, Annie Hayes, Alice Touchette, Anna White, and Donnell Oakley. I also wish to thank Noah Heringman, Kristen Pond, Julie Park, Jessica Damián, Kathryn Hansen, Rivka Swenson, Michelle Kohler, Michael Kuczynski, Rebecca Mark, Barry Ahearn, Supriya Nair, Scott Oldenburg, Adam McKeown, Molly Travis, Felipe Smith, T. R. Johnson, and Joel Dinerstein for creating environments of scholarly fun, kindness, and generosity. I feel very lucky in my editor, Angie Hogan, who believed in the project and conducted its publication in such a smooth, timely, and cheerful manner, and I appreciate the thoughtful readings and suggestions of the press's anonymous reviewers. Finally, and most especially, my heartfelt thanks go to my family, Sue, Eric, Katie, Alan, and Virginia Bailes, for their enduring faith, patience, and love, and to the memory of my father, Richard H. Bailes, who would have been proud to see this book in print.

Introduction

QUESTIONING SCIENCE, QUESTIONING ORIGINALITY

Questioning Nature argues that major women writers of the late eighteenth and early nineteenth centuries reconfigured literary, social, and national orders and identities through the natural sciences, achieving intertextual modes of originality that competed with now-conventional ideas of solitary Romantic genius. In mid-eighteenth-century Britain, rising fascination with developing disciplines of the biological and earth sciences during what has become known as "the second scientific revolution" coincided with literary critics' and authors' despair over the seeming paucity of poetic originality, the lack of new ideas for imaginative expression.[1] Responding to these critical anxieties about poetic novelty's possible extinction, the women of my study advanced scientific classifications, explorations, and discoveries in their literary works, contributing personal observations and theories, and attaining innovation by incorporating new scientific information. Such collaborative, scientific modes of writing complexly interacted with developing ideas of autonomous authorship. *Questioning Nature* thus explains how these women writers' imaginative scientific works both shaped the literary canon and led to their exclusion from it. Unlike any other book-length study, this project explores the connections between this era's phenomenon of women's scientific writing and their concerns with authorship, authority, and novelty, unveiling a new genealogy of Romantic originality.

At a time when women did not possess political power, some of the era's most prominent female authors exerted cultural authority through the natural sciences, particularly natural history (comprising the fields of botany, zoology, and geology). During this time, naturalists drew analogies between natural and social orders, arranging "classes" and "kingdoms" in ways that naturalized cultural and national hierarchies. As *Questioning Nature* demonstrates, women writers, including Anna Barbauld, Charlotte Smith, Helen Maria Williams, and Mary Shelley, appropriated this paradigm, claiming scientific superiority to naturalists to destabilize seemingly fixed identities and reconstruct not only biological but also social and literary taxonomies. For example, the travel writer and poet Maria Riddell used naturalists' theories of race and zoological hybridity to gesture toward West Indian colonies' potential to foster the hybrid British national identity, while Anna Seward devised a literary taxonomy that interrelates biological and poetic "species."

Surveying a complicated history of science from which women traditionally were barred, this book examines a mode of literary originality characterized by collective collaboration. As scholars such as Robert Macfarlane have shown, although male Romantic poets, including William Wordsworth and Percy Shelley, described poetic originality in wide-ranging terms, spanning a spectrum from reliance on multiple sources to creative generation solely in the poet's mind, it was literary critics of the 1820s and '30s who selectively championed the enduring and retrospective myth of the solitary Romantic genius.[2] In doing so, as I demonstrate, these critics also rewrote and condemned the roles of women and science in this era's literary history. I thus contend that to understand these male Romantics' conflicted claims and critics' historical influence it is necessary first to investigate this earlier movement of women who combined science and literature, producing hybrid forms of originality that challenged and redefined ideas of poetic autonomy.

The women writers of my project often cited or contextualized their scientific theories and observations in relation to those of established male naturalists, paradoxically realizing originality through intertextuality in forms such as plagiarism and borrowing, footnotes and endnotes, and reference and translation. Scientific texts themselves generally represented collective, collaborative undertakings, building on, referencing, and correcting other naturalists' systems, theories, and observations. These women writers, too, often question, expand, or confirm naturalists' claims. Even as these women strive for originality, their textual hybridities represent creative and cultural concerns as well as productive possibilities that interrogate the idea of isolated genius and emphasize competing networks of authorship and of scientific knowledge.

Largely adhering to chronology, each of this project's chapters analyzes how an individual female author uniquely renovates notions of originality through natural history while engaging with not only scientific matters but also some of the most important literary, racial, or national questions of her day. Anna Barbauld, for instance, debates critical differences between scientific verse and prose, exploring the extent to which natural history could stimulate poetic novelty. Maria Riddell, for her part, helps reveal a missing link in the georgic's scientific legacy of combining science and literature, highlighting zoological and textual hybridity as originality while examining colonial and national origins and values. Likewise viewing science as a model for literature's potential originality, Anna Seward and Charlotte Smith each compare biological with poetic species and clash over critical standards of plagiarism and imitation. And Helen Maria Williams infuses literary originality into her translations of scientific texts by integrating her own natural history poems and observations, often containing political implications. Participating in geological contentions about species' origins and extinctions, Mary Shelley then registers late-Romantic shifts toward critical preferences for autonomous authorship, complicating this scientific, collaborative mode of novelty through alterations that become explicit in Felicia Hemans's published and unpublished geological poetry. A close study of these writers, I argue, calls for new narratives of Romanticism, reflecting this literature's broad, ideological conflicts as well as the roles of science and women in challenging and creating now-canonical concepts of Romantic originality.

Modern critics have offered useful insights into how gender relates to natural history in the Romantic period yet overlook the significance of women's scientific literature for contemporary ideas of authorship. Extending the groundbreaking work of such scholars as Ann Shteir and Sam George in botany, Londa Schiebinger and Alan Bewell in zoology, and Noah Heringman and Ralph O'Connor in geology, this book provides a cultural history of the natural sciences in this era, while adding focus on women's goals in reformulating frameworks of social, biological, and literary hybridity and originality.[3] Moreover, these women writers arguably complicate what Gillian Russell and Clara Tuite term "Romantic sociability," which also contradicts the concept of the solitary Romantic genius and seeks to recognize the "fluid interplay" of literary circles "with other modes of sociability within British society as a whole."[4] Although I attend to sociable circles in which these individual women authors gained or received influence, my particular concern with science and intertextuality draws out authorial identities that employ and critique textual interchange while acknowledging various experiences of rejection and isolation. Indeed, women's scientific appropriations established such strong associations between women and natural history in the public imagination that not only did Wordsworth and other male Romantics actively seek to separate their verse from this movement of scientific literature but also, by the early nineteenth century, male scientific professionals diligently labored to reinstate the masculinity of studying the natural sciences.[5]

Discussing naturalists' systems, methodologies, and early notions of evolution, women reconceived literary novelty and attained social leverage in the public sphere that did not necessarily appeal to male authors in the same way.[6] As Susan Wolfson delineates, in this period any woman negotiated "a hot zone if her writing advanced the new philosophies of rights and liberties. The temperate zone hosted the home genres: conduct, sentiment, children, cookery."[7] In their scientific literature, women often blended these "hot" and "temperate" zones. While concern with poetry draws together each of my chapters, I examine women's treatment of natural history in numerous additional genres, including novels, travel writing, children's literature, translations, literary criticism, and periodicals. Scholars such as Anne Mellor have demonstrated that female authors of this age could win influence through discourses of moral conviction (for instance, in antislavery campaigns) and Republican motherhood (as in proposals for education); women also constructively combined these lexicons of female authority with that of science.[8] Although my study primarily pursues such manifestations of scientific discovery and authority in British literature and culture, I additionally follow these women writers' physical and imaginative travels to the West Indian colonies, to continental Europe, to the South Pacific, and to Central and South America, among other familiar and exotic locations in their quests for novelty. Crucial to understanding natural history's appeal for these female authors, as well as their resulting interactions with naturalists and tensions with some male Romantic poets, is contextualization within contemporary controversies about poetic originality.

ORIGINALITY IN CRISIS: LITERARY PLAGIARISM AND IMITATION

In the mid-eighteenth century, British literary critics and authors anxiously portrayed poetry in a state of crisis. Complaining that modern poets only plagiarized, both from classical authors as well as one another, these writers feared poetry's complete exhaustion, claiming that no new subjects for verse remained. In 1753 Samuel Johnson explained, "It is often charged upon writers, that with all their pretensions to genius and discoveries, they do little more than copy one another," and this "allegation of resemblance between authors is indisputably true."[9] A few years later, Oliver Goldsmith confirmed the nation's "degeneracy," linking Britain's literary decay with its moral decline.[10] In contrast to these modern failings, critics and writers frequently viewed their counterparts in the past, and especially classical authors such as Homer, as representing literature in its greatest state of perfection. Classical writers' superiority, it was thought, owed largely to their chronological priority, their seizure of all subjects open to them in nature, leaving modern authors little choice but to imitate their predecessors.[11]

Recognition of their literary progenitors' greatness and the difficulty of living up to this legacy weighed on the minds of mid-eighteenth-century authors, creating what Walter Jackson Bate termed "the burden of the past."[12] As Addison stated in the *Spectator,* paraphrasing Boileau, "It is impossible for us, who live in the latter ages of the world to make observations . . . which have not been touched upon by others."[13] In the *Guardian,* Richard Steele complains, "Nature being still the same, it is impossible for any modern writer to paint her otherwise than the ancients have done."[14] By midcentury, Joseph Warton only echoes that contemporary writers, "though surrounded with such a multitude of novelties, would find it difficult or impossible to be totally original, and essentially different from Homer and Sophocles."[15] Critical debates erupted, blaming or defending Restoration- and Augustan-era poets, especially Dryden, Pope, and Swift, for popularizing imitative verse. Outlining these perceptions of literary borrowings, Richard Terry explains that, earlier in the eighteenth century, sometimes "the reliance of modern authors purely on their own creative reserves, represented a form of cultural insularity and chauvinism" so that imitation became valorized.[16] For example, Swift memorably devalues literary originality in his fable of the spider and the bee, in *Battle of the Books* (1704), championing the bee's foraging through flowers of the field, and condemning the "surly, self-sufficient" spider, spinning his web out of his own entrails. Instead of this entomological analogy, the comparison of birds and bards would become especially significant in discussions about plagiarism and originality for both Anna Seward and Charlotte Smith, as it was for Pope, known as the Swan of Twickenham. In his *Peri Bathous* (1727), Pope classifies plagiarizing poets as "parrots," and his detractors similarly identified him, in "An Epistle to the Egregious Mr. Pope," as stealing others' plumage, a "Jay in Peacock's Beauties plum'd," based on a fable that Smith would later employ in her own defense against accusations of literary borrowing. After Pope's death, his use of literary models, borrowings, and imitations became critically suspect as servile plagiarism, helping to signal a shift in views of authorial property.

Whereas earlier in the eighteenth century plagiarism represented a moral shortcoming, by the mid-eighteenth century imitation and plagiarism became more closely aligned, typically drawing aesthetic rather than ethical indictments.[17] The turn against imitation is a marked feature of literary criticism in the 1750s. While the seriousness attached to charges of imitation and plagiarism caused many critics and authors to emphasize circumspection in leveling such accusations, there also thrived widespread critical exposure of potential poetical plagiarisms in what Macfarlane calls "the evolution of the 'plagiarism hunter'" comprising "a species of literary journalist which specialized in tracking down allusions, borrowings, and derivations, and then in listing these examples in an article as an arraignment of an author's originality."[18]

Despite their attacks on poetical copying, most midcentury critics affirmed faith in nature, the human mind, or both, to provide literary novelty. In this vein, Johnson exults that "the mutability of mankind will always furnish writers with new images" and, in his *Inquiry into the Nature and Genuine Laws of Poetry* (1778), Percival Stockdale suggests poets' ability to avoid plagiarism by "copy[ing] the ample, and inexhaustible page of nature."[19] Other critics privileged the imagination, founding this hope for novelty on ideas of the poet as "true genius." Often cited as a catalyst for Romantic-era notions of genius, Edward Young's *Conjectures on Original Composition* (1759) glorifies poetic originality as not only copying nature rather than art, but also as "partak[ing] of something Divine," of God's generative powers, famously lamenting, "born Originals, how comes it to pass that we die copies?"[20] Young, Goldsmith, and Stockdale each assert that certain nations prove more conducive to this poetic talent than others, adopting contemporary naturalists' theories that climate and soil affect biological species and extending those effects to the production of great literature. Stockdale, for example, patriotically locates poetic genius in Britain's climate, remarking "that no education would enable a native of the Equator, or of Greenland to distinguish himself by his mental faculties."[21] Nevertheless, while proclaiming confidence in the ability of poetic genius to find new subjects for verse, critics offered few specific answers to this problem of producing originality. One solution, however, brought together these ideas of directly copying nature and of substantiating social and poetic assertions through scientific thought, advocating a kind of natural history poetry that would shape the course of literary history.

Scientific Originality

In their pursuit of literary originality, critics and authors increasingly championed new discoveries made during "the second scientific revolution." Between the mid-eighteenth and mid-nineteenth centuries, scientific advancements particularly thrived in natural history. Critics such as Edward Young, Oliver Goldsmith, Percival Stockdale, and Thomas Percival each gestured toward melding literature and science as a means to poetic originality, and John Aikin arguably articulated its clearest espousal in his *Essay on the Application of Natural History to Poetry* (1777).[22] At this time, naturalists racing to describe and classify biological organisms called for the assistance of amateur observers whose knowledge of local ecologies could help map the natural order. Echoing this call to nonspecialists, Aikin encouraged poets to further scientific endeavors through study and close observation of nature both to assist naturalists and to reveal original subjects for verse. As my examination shows, his essay complexly appealed to female poets, particularly Barbauld and Smith. Although traditionally composed in Latin, natural history texts increasingly were written or translated in English, and considerably rose in popularity in the mid-eighteenth century and succeeding decades as they targeted a wider audience, including women. Indeed, during this era, the British botanist Peter Collinson enthused that natural history sold "the *best of any books* in England."[23]

While natural histories, especially those containing illustrative plates, could be exorbitantly expensive, the affordable price of periodicals helped disseminate scientific knowledge to an even broader public. Often incorporating summaries of naturalists' latest arguments, systems, and discoveries, periodicals provided a venue for amateur contributors to voice their thoughts about these findings, and sometimes promoted science's specific attractions for women.[24] For example, two of the first periodicals written by and for women, Eliza Haywood's *Female Spectator* (1744–46) and Charlotte Lennox's *Lady's Museum* (1760–61), advocated women's study of natural history in empowering terms.[25]

Haywood's periodical persona and her (probably fictional) male contributor, Philo-Naturae, as well as, later, Lennox's persona, each blends instruction and amusement while boldly suggesting that women's knowledge of natural history could surpass that of men, enabling women, for instance, to make discoveries "not to be found in the most accurate Volumes of Natural Philosophy" so that "the *Royal Society* might be indebted to every fair *Columbus* for a new World of Beings to employ their Speculations."[26] Assuring female readers that scientific proficiency may be attained in a single summer of study, Haywood encourages them to carry miniature microscopes to analyze plants and insects encountered during excursions as well as to conduct experiments, testing scientific theories rather than taking naturalists at their word.[27] She shrewdly portrays women's scientific supremacy as inseparable from feminine propriety, frequently alluding to science as a means of religious devotion.[28] As Philo-Naturae states, "the study of *Nature* is the study of *Divinity*," and the Female Spectator affirms that "a sincere and ardent love to God would be conveyed to us through our admiration of his works."[29] Most eighteenth-century naturalists highlighted their participation in natural theology, a conventional scientific piety that helped facilitate women's access to natural history, and texts by Mary Shelley and Felicia Hemans demonstrate how this access became less assured with science's increasing secularization in the early nineteenth century.

During the second scientific revolution, simultaneously collaborative and competitive generations of prominent European naturalists captivated Western thought, striving to configure the order of the natural world. The Swede Carl Linnaeus taxonomized zoological and botanical species, becoming especially famous for his "sexual system" of botany that classed plants according to their sexual parts. In France, Georges-Louis Leclerc, comte de Buffon, combined interests in zoology and geology, espousing theories about the effects of soil and climate on biological species. His younger countryman Georges Cuvier also associated zoology and geology in his studies of paleontology and comparative anatomy to prove the occurrence of species extinction through reconstructed fossils and past eras of the earth. In Britain, Joseph Banks, who accompanied Captain James Cook's first voyage, discovering many new botanical and zoological species and representing the expanse of Britain's global grasp of natural history, became president of the Royal Society, while other naturalists, such as Thomas Pennant and Gilbert White, more strongly stressed local knowledge of British nature. The

German Alexander von Humboldt then conjoined this attention to local and universal nature in his grand system now known as Humboldtian science. These celebrated naturalists' disparate methodologies engendered interest from women writers who uniquely applied science's social and literary authority.

Indeed, inquiries in the natural sciences often had sociological implications. Naturalists investigated the extent to which biological organisms, including humans, shape and are shaped by their environment, and assessed nature as a model for society, while also imposing social order onto nature. Naturalists thus classified and defined not only natural objects, but also human subjects, formulating (often hierarchical) theories of gender, race, and nation through interpretations of nature. By appropriating naturalists' social analogies, imaginative authors employed botany, for example, as a means of exploring gender as well as class and sexuality, zoology as a discourse for race, nation, and hybridity, and geology as a framework for political revolution, beginnings and endings, and the geohistorical sublime. Moreover, through their literature, these women writers participated in scientific controversies about, for example, whether biological species were fixed and unalterable or contained possibilities for transformation and change. Such questions had implications for ideas of originality and hybridity, concepts that could be viewed as cooperative and virtually interchangeable by authors including Riddell, or as opposing models of biological (and literary) forms, in which hybridity represents monstrosity, as they were by Seward. In this way, writers engaged with natural history debates to substantiate their separate understandings of literary and social purpose, meaning, function, and construction. In their authoritative knowledge of nature, and goals to find or create order, naturalists influenced perceptions of every known level of earthly existence and, through science, imaginative writers could adopt this powerful stance.

Despite these commanding claims to knowledge of the natural order, naturalists' observations, classifications, and theories were subject to incessant attack and revision. Linnaeus and Buffon, for instance, ruthlessly besieged one another's published methods and taxonomic theories throughout their careers.[30] Naturalists also corrected their own work in successive editions as new information or discoveries became available. Science's resulting sense of flux put the line between fact and fiction in doubt, blurring the division between literature and natural history. Of course, in the eighteenth century the sciences were still in development and, although I use the word "science" in its modern sense, previous to science's professionalization the term could constitute any branch of knowledge, including literature.[31] Through this disciplinary fluidity, imaginative writers could contribute to scientific knowledge in their literary works. Natural history's simultaneous authority and uncertainty, stemming from openness to inquiry, challenge, and debate, created opportunities for women writers' scientific participation, especially in poetry.

RETHINKING THE GEORGIC INFLUENCE

Romantic-era poets versifying new scientific findings and ideas in the manner advised by Aikin and other critics sometimes highlight influence from the well-established georgic mode. The phenomenon of the "English georgic" principally occurred between 1708 and 1767, following Dryden's published translation of Virgil's *Georgics* (1697).[32] Blending didactic and descriptive styles, the georgic conventionally emphasizes expansive prospects and intensive agrarian labor toward the improvement, civilization, and building of a nation. Poems considered "true" georgics, securing the genre's authority, include John Philips's *Cyder* (1708), Pope's *Windsor-Forest* (1713), James Thomson's *The Seasons* (1726–44), William Somervile's *The Chace* (1735), Christopher Smart's *The Hop-Garden* (1752), John Dyer's *The Fleece* (1757), and Richard Jago's *Edge-Hill* (1767). Speculating about the fate of the georgic, scholars variously attribute the form's seeming disappearance after 1767 to the Industrial Revolution, to displacement of scientific agricultural discourse from poetry into prose treatises, to Britain's loss of the American war and the diversion of agrarian ideals from George III to the newly constituted United States, to what Clifford Siskin calls the "lyric turn" that devalued georgic garrulousness and pedagogical purpose in favor of a more compact form of immediacy and imaginative vision, and to a transformed and enduring legacy in poetry such as William Cowper's *The Task* and Wordsworth's *The Excursion*.[33] Significantly, the georgic's disappearance also functions as a dissemination that may be traced into the scientific literature of this study.

The English georgic's associations with agrarian labor, technical science, nationalism, and politics define it as a traditionally masculine poetic form. Yet as critical favor gradually shifted from didacticism to description, women writers engaged with and reframed georgic verse, helping to guide its fate by transforming it through their own works. Women such as Maria Riddell, Anna Seward, and Charlotte Smith singled out Thomson's *The Seasons,* the eighteenth century's most popular georgic, as a model against which to demonstrate their patriotic contributions and superior scientific knowledge, building on his critical acclaim. In his *Essay on the Application of Natural History to Poetry,* Aikin specifically praises Thomson as "the Naturalist's Poet" for his close observations of nature that exemplify ways for poets to become better, more knowledgeable naturalists than the naturalists themselves.[34] According to Aikin, poetry's beauty depends on its truth, its accuracy of description. For him, Thomson's descriptive emphasis gives his georgic supremacy even over that of Virgil, who places more weight on didacticism. Aikin thus lauds "the result of Thomson's unconfined plan, scarcely less extensive than nature itself" while lamenting that "some other writers, not inferior in genius . . . thought it necessary to shackle themselves with teaching an art or inculcating a system."[35] Often staking claims to greater technical authority than Thomson, the ultimate poet-naturalist, women writers positioned themselves as inheritors of this scientific poetry, which they could appropriate and negotiate in its dual emphasis on pedagogy and pleasure. Although critically deemed "didactic poetry," the georgic's tension between description and didacticism incited controversies over how the form should manifest in subsequent scientific verse.

While Goldsmith asserted that "the English are deservedly famous" for didactic poetry, other critics voiced concern about the form's educative purpose.[36] In his *Letters of Literature* (1785), John Pinkerton, under the pseudonym Robert Heron, denounces didactic verse as "the lowest of all kinds of poetry" because it allows for little "invention," forcing poets to write pedagogically.[37] Underlying this complaint is the idea that technical discussion and detail ruins poetry's pleasure. In this vein, Stockdale agrees that "if the subject of a Poem is obscure, or not generally known, or not interesting, and if it abounds with allusions, and facts of this improper, and uninteresting character, the writer who chuses that subject . . . betrays a great want of poetical judgement, and taste."[38] In other words, poetry is not the proper venue for pedagogy, or for learning anything new. Moreover, the explanatory notes required in didactic verse ensure "the flow and warmth of the reader's mind, while He accompanies the poet, is checked and broken," for, according to Stockdale, poetry should not demand exertion of one's "*understanding.*"[39] As

Kevis Goodman points out, this georgic conflict "between the technical and poetical is not just the Horatian commonplace about negotiating a harmonious balance between instruction and pleasure (utile and dulci)," but a genuine question of whether technical, scientific subjects are fit for poetry.[40] Of Dyer's georgic, *The Fleece,* about sheep-shearing and the cotton market, Johnson pronounced, "The subject, Sir, cannot be made poetical."[41] In contrast, Joseph Trapp wrote an often-reissued treatise, insisting, "Precepts and Poetry are no ways inconsistent" as long as the "*technical words,* or *Terms of Art* . . . might be so dressed up, as to invite, not deter the Pains of the Listener."[42] This divergence of opinions becomes a problem that Romantic-era female poet-naturalists seek to resolve, for while Charlotte Smith's verse incorporates erudite details, such as species' binomial nomenclature, Anna Barbauld circumspectly considers ways in which scientific instruction prohibits poetic pleasure.

In addition to this possibility that incorporation of new facts could ruin poetry, some critics warned against education's detrimental effects on the true poet. Writing in 1767, William Duff, for example, argues that early poets' lack of "Learning and critical Knowledge" allowed verse to reach "its highest perfection in the first uncultivated periods of human society."[43] He denigrates poets of descriptive scientific verse, explaining that "By collecting the observation and experience of past ages, by superadding his own, and by reasoning justly from acknowledged principles, [the poet] may, no doubt, acquire more accurate and extensive ideas of the works of Nature and Art, and may likewise be thereby qualified to inrich the Sciences with new discoveries" but, in poetry, "he can never become an original Author by such means."[44] For Duff, poets' attempts to make "new discoveries" in fields such as scientific knowledge can only hinder original poetry. Anticipating the trope of the solitary Romantic genius, he states that the poet "must rely on his own fund: his own plastic imagination must supply him with everything."[45] Nevertheless, the earlier, neoclassical concept of imitation also persists in this period as capable of both recapitulation and creation.[46] Contemporary science derived its authority from the replicability of observations and experiments, and some naturalists and imaginative authors perceived variation in the repetition of older (biological and literary) forms as a means to something new. Such critical debates about the extent to which poets may meld science and verse to engender poetic originality recur and evolve over the course of the Romantic era. Opposing Duff's critical view, most women writers in this study join "ideas derived from books" with their observations of nature specifically to make original literary and scientific contributions.[47]

A Different Kind of Romantic-Era Nature Poetry

Certainly, in this period, male as well as female writers combined natural history with literature.[48] In her essay "'Unsex'd Females': Barbauld, Robinson, and Smith," Judith Pascoe reproaches recent feminist criticism for reading women's literature in gendered terms that portray particular genres, such as sentimental animal poems, as "veiled critiques of masculine power structures" when, in fact, a number of male authors concurrently wrote in that genre.[49] I take Pascoe's caution seriously and examine male writers such as Thomson, James Grainger, Erasmus Darwin, Wordsworth, and Byron in their literary engagements with science. Yet women played a central role in this trend of literary naturalism. While representations of nature pervade Romantic-era literature, women writers often exhibit a more minutely detailed and taxonomic approach to natural objects than their male counterparts of "high" Romanticism, and they explicitly adopt scientific models, classifications, and concerns. Seeking to explain this particular window of time in which women responded to science through literature, recent scholars attribute its beginning to the rise of specific taxonomic systems, like that of Linnaeus, and its disappearance to science's professionalization in the early nineteenth century, resulting in the exclusion of women and amateur naturalists from serious participation in the field.[50] Expanding this traditional account, my project shows that changes in concepts of originality both affected this movement's involvement in sociobiological debates and crucially explain how and why such literature historically arose in the latter half of the eighteenth century and ended or altered around the 1830s.

These women authors and, later, the male Romantic poets such as Wordsworth, arguably wrote in reaction to the same problem, this critically unsettling crisis of literary novelty, but did so with different implications for ideas of originality. While women's merging of literature and science allowed them to make original contributions in both fields, often correcting naturalists through personal observations, discoveries, and hypotheses in the notes or body of their works, such literature, as previously stated, paradoxically achieves novelty by revealing its intertextuality. In doing so, this literature lends itself to those earlier, neoclassical ideas in which reference, collaboration, and improvement could constitute novelty, especially since, in the latter half of the eighteenth and early nineteenth centuries, theories of plagiarism, imitation, and originality were hotly debated. Numerous scholars have explored literary originality's instability in the Romantic period, exhibiting that even in the texts of those male poets now most prominently associated with solitary, imaginative creation, autonomous originality was an aesthetic fantasy, and those male writers in fact promulgated complex ideas about literature's freedom from outside influences.[51] For example, Wordsworth both called the imagination "the source from which everything primarily flows," and emphasized multitudinous sources of thought.[52] Likewise, in *A Defence of Poetry* (1820–21), Percy Shelley envisaged the creation of wholly "new materials" while, in *Prometheus Unbound* (1820), he described creativity only as combination and "analogy."[53] Eliding these inconsistencies and spectrums in determining original literary productions in the early decades of the nineteenth century, as Macfarlane has shown, literary critics began to simplify and mythify the notion of solitary poetic genius, emphasizing the works of Wordsworth and other male Romantics as constituting a new kind of poetry.

Early nineteenth-century critics thus portrayed a schism, championing unindebted originality as separate from intertextual kinds of literature. This gradual process of division and literary elitism employed selective hindsight, highlighting ways in which male Romantics sought originality by reacting against other literary movements, including that of these women writers who pursued originality through science. For example, in his *Lectures on the English Poets* (1818), William Hazlitt separates Wordsworth from such intertextual scientific poetry, explaining that this male poet "tolerates only what he himself creates" with his "power and originality of mind" and therefore "hates all science and art; he hates chemistry, he hates conchology."[54] According to him, Wordsworth "sympathizes only with what can enter into no competition with him"[55] Critics such as Catherine Ross have analyzed Wordsworth's intense feelings of competition with the chemistry of Sir Humphry Davy, and Hazlitt's inclusion of conchology, a popular branch of natural history, as another science this poet "hates" gestures toward Wordsworth's desire also to distance his poetry from subjects that had become so associated with these women writers.[56] Hazlitt greatly influenced the shaping of the British canon, and his commentary on Wordsworth, purging science, and women, demonstrates subsequent consequences for authors in this scientific movement. Such critical exclusions super-seded earlier attempts at canon making, like that of Anna Seward, who here indirectly becomes precluded from the literary taxonomy in which she strove to find a place. Later, in 1848, as Tim Fulford, Debbie Lee, and Peter Kitson explain, Thomas De Quincey "made fiction a defin-ing characteristic of the 'literature of power' and claimed it was distinct from the 'literature of knowledge,'" thereby "fenc[ing] off travel writ-ing, natural history, political journalism, to name but a few genres, from the realm of 'high' literature—that which communicated across time, through the aesthetic mode of the sublime."[57] Clearly, in this distinction, De Quincey's "high" "literature of power" describes the mode of imag-inative writing now correlated with male poets of Romanticism, such as Wordsworth, Percy Shelley, and Byron. By contrast, the women's literature I examine, although often encompassing imaginative genres, makes a point of conveying scientific "knowledge." Yet, in the late eigh-teenth and early nineteenth centuries, these two kinds of literature vied for prominence.

While, for a time, science and literature seemed compatible and in-terchangeable sources of authority, some male Romantics, writing after this women's movement of scientific literature was established, pur-sued alternative modes of literary originality and imaginative concepts of nature. Indeed, I discuss how Wordsworth's preface to *Lyrical Bal-lads* distinguishes poetry from science and how Byron employs natural history for satirical effect, and Theresa Kelley has displayed that even John Clare, who writes with a naturalist's attention to nature, often re-sists technical, scientific language and thought.[58] In contrast, although women writers in my project sometimes held goals in opposition to one another, they viewed literary incorporation of scientific authority as a means to make things happen, to contribute new knowledge and obser-vations, to participate in scientific and national debates, to challenge or uphold established social and literary identities and agendas, as well as to attain novelty.

Recent scholars have noted that other male contemporaries, such as Erasmus Darwin, wrote scientific poetry more closely resembling that of these female authors, especially in structure, themes, and cultural concerns.[59] While several women writers responded to Darwin's work, at the same time, as I show elsewhere in this text, scientific literature by women, including Anna Barbauld, Anna Seward, and Charlotte Smith, predates his scientific poetry so that his verses' influence should not be overstated in motivating women's literary engagements with natural history. Interestingly, Darwin and other male writers who specifically address natural history repeatedly do so in terms that bring the focus back to women by feminizing nature and science, by defining and metaphorically confining women, or by discussing science as a subject of women's study.[60] In one way or another, women persistently became bound up in conversations of natural history. Indeed, this feminization of natural history sometimes played a role in contemporary lampoons of science, and women writers, such as Barbauld and Hemans, also could display humor in their scientific subjects.[61] Concurrently with their reframing of Enlightenment science and eighteenth-century notions of originality, these women selectively deployed Augustan satire and "mock" styles to attain poetic distance and superiority in their authoritative critiques of naturalists and natural history, opening a space for their own serious contributions to scientific thought.

CATEGORIES OF INQUIRY: CHAPTER SUMMARIES

Many additional women could have been included in this study of Romantic-era female writers who employed natural history to achieve literary originality. A single book cannot hope to cover all of the personalities involved, or their various ways of contributing to such a movement, and Mary Wollstonecraft, Priscilla Wakefield, Dorothy Wordsworth, Maria Edgeworth, Sarah Hoare, and Maria Graham represent only a few further female authors who do not receive chapters here, but merit attention within such a project. Nevertheless, each of my chapters analyzes a specific woman author whose ideas about origins and originality helps map the trajectory of changes in literary and scientific thought during this span of time. Supporting the chronological progression of my study, the women writers I chose display the historically shifting levels of interest and available information in the respective fields of botany, zoology, and geology, and thereby highlight different naturalists whose ideas influenced this literary movement. In addition to demonstrating the broad range of intertextual literary forms employed by women engaging with science, each of these authors also interrogated what constitutes originality, exhibiting distinct reasons for viewing natural history as the means to literary novelty.

As showcased by the book's organizing scheme, its chapters pair into three main sections. The first two chapters treat Anna Barbauld and

Maria Riddell as case studies of natural history's literary associations with gender and nationalism, displaying these authors' differing theories about what should be included in original scientific literature, and why they made such decisions as this literary movement took shape. The next two chapters, forming the book's core, focus on the poetic rivalry between Anna Seward and Charlotte Smith, exploring how natural history's potential for novelty interacted with ideas of literary plagiarism and hybridity, as well as solitude and sociability. The final two chapters demonstrate ways in which texts by Helen Maria Williams and Mary Shelley challenge scientific hypotheses connecting political revolution and the geological sciences, building on earlier chapters to reveal why this literary movement failed or evolved. Over the course of six chapters, I thus trace the renovating possibilities natural history held for female authors as well as how, after the 1820s and '30s, it became impossible for women to practice scientific literature in the same way.

Chapter 1 examines Anna Barbauld's opposing theories about incorporating scientific information into verse and into prose, comparing her natural history poetry (1773) with her pedagogical works for children (1778, 1781), especially in relation to John Aikin's *An Essay on the Application of Natural History to Poetry* (1777). Aikin asserts that poets should study the natural sciences to discover original subjects for verse, and he invites women poets' participation, foregrounding the verses of his sister, Barbauld. However, despite this gesture, Barbauld represents a conflicted image of the poet-naturalist, for although she attributes novelty to natural history poetry, she also proposes that poetic use of science should extend only to natural objects and phenomena already familiar within the public imagination. She thereby anticipates Wordsworth's critique of scientific poetry later expressed in the preface to *Lyrical Ballads* (1800). While Barbauld highlights Caribbean plants as exemplifying these unfamiliar and therefore inappropriate specimens for versification, it is precisely to the West Indies that my second chapter turns. Chapter 2 focuses on the transatlantic travel literature (1792) of the poet Maria Riddell and her appropriation of the georgic's nation-building power. Her travel writing combines science and poetry, attaining originality while interrogating ideas of literary, national, and biological degeneration and improvement. I claim that Riddell melds ideas of origins and originality, drawing on the georgic's formal and national hybridities and on naturalists' theories of zoological species' intermixing, to present West Indian colonists of Scottish, Welsh, Irish, and English origin as hybrid "Britons" in a now-shared experience of common values and a broader sense of national (rather than regional) allegiances.

Scientific conceptions of hybridity also inform my third and fourth chapters, exploring poetic originality within the context of Anna Seward's famous accusations of plagiarism against rival poet Charlotte Smith (1784, 1804). Chapter 3 asserts that Seward's thinking about poetic imitation was shaped by a belief in fixed biological forms that tended to see newness as divisible into two categories: those of originality and hybridity. Distrusting deviations from originality, Seward considered Smith's and Erasmus Darwin's poetic plagiarisms to be degenerate, stylistic hybrids and precluded them from her literary taxonomy, constructed to reflect the natural order. Chapter 4, in turn, takes up this question of science's relation to literary originality from the perspective of Charlotte Smith, especially examining her late texts, *Conversations Introducing Poetry: Chiefly on Subjects of Natural History* (1804), *The Natural History of Birds* (1807), and *Beachy Head, Fables, and Other Poems* (1807). Punning on natural history's practice of collecting, Smith's poetic borrowings achieve what I term "collective originality," yet her efforts to produce original works finally lead to her troubling realization that the copying of *nature* may just as easily draw accusations of plagiarism as the copying of *art.*

Smith's eventual skepticism about natural history as a means to literary authority and originality parallels Helen Maria Williams's late doubts about science in the political realm. Chapter 5 analyzes how Williams employs concepts of nature and nation to produce literary novelty in a surprising medium: her translations of scientific texts. In addition to versifying the death and legacy of Captain James Cook (1788), whose voyages popularized the natural history of the South Pacific, she translated naturalists' works, such as J. H. Bernardin Saint-Pierre's *Paul and Virginia* (1795), Louis Ramond de Carbonnières's essay on Alpine glaciers (1798), and Alexander von Humboldt's explorations of Latin and South America (1814–29). Williams interpolates her own original poetry and scientific footnotes and alters naturalists' diction in her translations, creating originality and shaping the way British readers received these scientific ideas. In doing so, she also illustrates connections between her changing perceptions of natural history and the political climate in France.

Chapter 6 demonstrates how correlations between political and geological revolutions, emphasized by Williams, differently manifest in literary depictions of the global future. Here, I argue that Mary Shelley's *The Last Man* (1826) critiques paleontological theories about what caused past species extinctions, ultimately endorsing that of plague. Juxtaposing extinction hypotheses as a means of thinking about literary and biological origins as well as endings, she shifts Georges Cuvier's idea of geological catastrophism into the psychological "world" of the individual. Revising literary techniques employed by the women writers discussed in previous chapters, Shelley also engages science in stylistic terms and concepts of originality conventionally associated with the male Romantics, signaling alterations in women's scientific literature. Although, as my final chapters show, these effects had been in motion for some time, in the 1820s developing Victorian ideals of feminine propriety combined with the increasing secularization of science as well as the professionalizations of literature and natural history to make it more difficult for women to participate in serious scientific discussion, let alone to posit new scientific thoughts, observations, or discoveries through imaginative literature. Additionally, at this time, critical expectations of originality shifted more prominently toward the championing of autonomous genius, rather than collaborative forms. My conclusion analyzes Felicia Hemans's geological poems (1816, 1828) as responding to these cultural conditions. Her poetry exemplifies, I claim, an alternate, domestically oriented literary and social potential that incorporates now-established tropes of Romantic imagination and originality and sets the new standard for women's scientific literature in succeeding decades.

Gender and Nationalism

DESCRIBING AND DEFINING LITERARY NATURALISM

1

To Teach and to Please

ANNA BARBAULD'S ORIGINAL POETRY AND EDUCATIONAL PROSE OF NATURAL HISTORY

In a letter to Samuel Coleridge, Charles Lamb vehemently censures authors of contemporary educational texts, exhorting his friend to "think what you would have been now, if instead of being fed with Tales and old wives fables in childhood, you had been crammed with Geography & Natural History.? *Damn them.* I mean the cursed Barbauld Crew, those *Blights & Blasts* of all that is *Human* in man & child."[1] Opposing imaginative tales to cold, hard facts of reason, Lamb condemns Anna Barbauld for propounding "soul-killing rationalism."[2] He attacks natural history's encroachments on bulwarks of the imagination, lamenting that "Science has succeeded to Poetry no less in the little walks of Children than with Men." Yet, while Barbauld certainly encouraged rational inquiry into the natural sciences—for example, in her epistles "On Female Studies," stating that natural history comprises a discipline it is "unpardonable not to know"—her interest in such study, particularly for women, facilitated a more complicated negotiation of literature and science than Lamb's "curse[s]" suggest.[3]

The earliest of the women writers I address in this study, Barbauld registers natural history's empowering possibilities for women and literature of the Romantic era. She theorized where and how the blending of science and literature should occur, seeking balanced and effective strategies for women's original, imaginative works. Especially interrogating scientific verse, Barbauld anticipates and later revises assertions made by her brother, John Aikin, in his important *Essay on the Application of Natural History to Poetry* (1777). Although she agrees with Aikin that originality in contemporary poetry requires the natural sciences, she confines poets only to familiar aspects of these sciences. Moreover, she draws a distinction, contending that the purpose of poetry is to please and that of prose is to teach. Her engagements with natural history thus do not produce the simple hierarchy denounced by Lamb as privileging rational thinking over imaginative pleasures. Rather, pleasure vitally informs Barbauld's scientific verse and her advocacy of women's writing, as well as her distinction between prose and poetry. Critically formulating the extent to which science and literature should overlap, she designates prose as the appropriate venue for education while determining that poetry's pleasure and novelty depend on a more qualified approach.

Definition versus Description: Educational Prose versus Pleasurable Poetry

To understand Barbauld's theoretical distinction between scientific prose and poetry, it is useful first to turn to one of her pedagogical essays, exactly the sort that provoked Lamb's damning critique. Her "Lesson in the Art of Distinguishing" constitutes one of fourteen pieces she contributed to *Evenings at Home* (1792–96), a collection of stories for children, chiefly authored by her brother, John Aikin. Employing prose dialogue, one of Barbauld's favorite literary forms, her "Lesson" entails an edifying exchange between a young boy, Charles, and his father.[4] Here, by posing questions designed to determine exact differences between a horse and a number of other creatures, including a cow, cabbage, and salmon, Father leads Charles to a naturalist's precise definition: "A horse is an animal of the quadruped kind, whole-hoofed, with short erect ears, a flowing mane, and a tail covered in every part with long hairs. . . . [H]e has six cutting teeth in each jaw."[5] For Barbauld, the purpose of this exercise, as delineated by Charles's father, is the acquirement of rational skills: "I have not given you a definition to teach you what a horse is, but to teach you to *think,*" for "nothing is more useful than to learn to form ideas with precision and to express them with accuracy."[6] It seems to go without saying that a later Charles (Dickens, that is) almost certainly parodies this exchange in his novel *Hard Times* (1854), where the definition of a horse is famously demanded by the "square finger" of Gradgrind in a story that satirizes exaction of hard facts and reasoning from children at the expense of imagination.

Although Barbauld's essay most obviously demonstrates how to "distinguish" between biological species, it points to generic distinctions as well. Unsatisfied with the naturalist's scientific definition of a horse, the young Charles instead prefers to "say it was a fine large prancing creature, with slender legs and an arched neck . . . and that he snorts and neighs very loud . . . and runs as swift as the wind."[7] He then quotes an eloquent verse depiction of a horse from Pope's translation of Homer, to which his father replies, "You have said very well; but this is not a *Definition*, it is a *Description*." Father informs Charles:

> A description is intended to give you a lively picture of an object, as if you saw it; it ought to be very full. A definition gives no picture to those who have not seen it; it rather tells you what its subject is not, than what it is, by giving you such clear specific marks, that it shall not be possible to confound it with any thing else; and hence it is of the greatest use in throwing things into classes. We have a great many beautiful descriptions from ancient authors so loosely worded that we cannot certainly tell what animals are meant by them; whereas if they had given us definitions, three lines would have ascertained their meaning.[8]

To this Charles replies, "I like a description best."

For Barbauld, this difference between "definition" and "description" marks her distinction between scientific prose and descriptive poetry, especially descriptive nature-poetry. The poetic description of the horse (potentially "so loosely worded that we cannot certainly tell" what animal is meant by it) is less precise but, by Charles's estimation, more pleasing, than the prose definition, so apt for a naturalist's system of classification. How far these genres appropriately might meld is a question with which Barbauld struggled and, indeed, one she posed to her brother, John Aikin, in reaction to his *Essay on the Application of Natural History to Poetry*.

Aikin's *Essay* sought to alter poetic composition in his day, advocating the combination of poetry and the natural sciences as the best means to show originality in modern verse. He deplores poets' contentedness merely to copy phraseology and natural descriptions from poetic predecessors, often promulgating false images of nature. To combat this stagnation, Aikin instead recommends closer imitation of nature itself through accurate observation. He claims that nature's inexhaustible resources for poetic subject matter ensure not only novelty, but also greater aesthetic and moral value, for "nothing can be really beautiful which has not truth for its basis."[9]

Aikin's essay establishes a tone of nationalism and inclusivity, regardless of class or gender. He dedicates his work to the patriotic British naturalist Thomas Pennant, who authored the popular natural history text *British Zoology* (1768–70) nine years earlier. In fact, Aikin helped revise this zoology, the preface of which encourages descriptive poets' knowledge of natural history.[10] According to Pennant, nature provides a vast store of "metaphors, allusions, or descriptions," while the poet lends "life and motion to every object."[11] In his *Essay*, Aikin hopes to further Pennant's cause, inspiring poets as "fellow-labourers" in the research of natural history. Significantly, in recommending the study of nature, Aikin carefully notes that the identity of the ideal "poet-naturalist" is not "confined to the adept in systems and proficient in names."[12] Rather, "it is intended to comprise every one who surveys natural objects with a searching and distinguishing eye; whether he consider them singly, or as part of a system, whether he call them by their trivial or learned appellations." This gesture of inclusion embraces "every one" wishing to study nature through personal observation. Not preferring species' scientific over common names, Aikin's negation of the need for Latin or reference to scientific taxonomies invites the participation of amateur naturalists, including women.

Advocating accurate observation, Aikin promises immense scientific authority to poets. Although he reveres Pennant as an exemplary naturalist, Aikin suggests that poets should not confine themselves too didactically to propounding any naturalist's particular system. Indeed, he subordinates naturalists' authority to that of the poet, stating that poets will, through observation of nature, often acquire a knowledge of natural history that surpasses that of the naturalists themselves. Further, Aikin represents this study of nature as a particularly virtuous pursuit, promoting moral instruction and sentimental subjects for poetry as, for instance, when a live decoy duck unknowingly attracts, and thus entraps, his fellows.[13] Analyzing poetic passages, Aikin gauges the successes and failures of numerous classical and modern poets in recording accurate perceptions of nature, citing the works of Homer, Virgil, Pliny, Milton, Gray, and, of course, Thomson. He particularly admires the correct natural descriptions in Thomson's *The Seasons,* hoping some "second Thomson" will convert Pennant's scientific definitions of biological species into verse. In this vein, it is interesting to note the direction of suggested influence. Should science be incorporated into literature, or literature into science? Most of Aikin's examples from the literary tradition display the former, so that a writer begins with a literary theme and briefly employs a natural history comparison to enliven his description. However, by recommending Pennant's prose as a basis for novel poetry, Aikin urges poets to ground their subject more firmly in science itself.

Significantly, Aikin singles out Barbauld's "To Mrs. P[riestley], With Some Drawings of Birds and Insects" (1773) to exemplify successful versification of natural occurrences in which poetry expounds on science. This poem addresses Barbauld's friend Mary, wife of the famous natural philosopher Joseph Priestley. Here, Aikin highlights Barbauld's description of the chrysalis stage of insect development, "the transformation of the caterpillar . . . to its butterfly state," and her incorporation of a literary allusion to Torquato Tasso's *Gerusalemme Liberata* (1581). Depicting the final phase of pupae maturation, Barbauld makes natural processes poetically noteworthy and even hauntingly, paradoxically surreal through adherence to reality. She portrays these insects immured, emphasizing the cocoon as a "tomb" from which transformed butterflies "burst" to "leave their sordid spoils" of earth behind when ascending to the "Ether" (ll. 78, 80, 84). These insects' progression from metaphorical death into (immortal) life dramatizes Christian tradition as well as spiritual connotations associated with butterflies in classical mythology.[14] Barbauld compares this apocalyptic resurrection with Tasso's trees, whose "lovely births disclose,/. . . a gay troop of damsels" (ll. 88–89). Yet she relegates Tasso's fanciful description to a supportive role, emphasizing instead actual processes in nature. Just so, Aikin later would state that "the most vivid imagination cannot paint to itself scenes of grandeur equal to those which cool science and demonstration offer to the enlightened mind. . . . The most faithful pencil . . . produces the noblest pictures."[15] In his *Essay*, Aikin thus exemplifies Barbauld as a model poet-naturalist, and he was not alone in praising her "faithful" treatment of the natural sciences. For instance, a critic from the *Monthly Review*, examining her 1773 volume of poetry, also lauded this particular poem, noting that "it abounds with hints of considerable knowledge in natural history, and is void of affectation and philosophic pomp."[16] Barbauld, too, recognized this undesirable potential "of affectation and philosophic pomp" in poems of natural history, prompting her to demand deeper consideration of how such verse should be configured.

When Aikin sent Barbauld a copy of his *Essay on the Application of Natural History to Poetry,* she questioned the consequences and scope of her brother's plan. While acceding "that the only chance we have for novelty is by a more accurate observation of the works of Nature," she also sought a more specific explanation of his literary theory.[17] In a letter to him, Barbauld cautions, "it would not have been amiss if you had drawn the line between the poet and natural historian; and shown how far, and in what cases, the one may avail himself of the knowledge of the other,—at what nice period that knowledge becomes so generally spread as to authorize the poetical describer to use it without shocking the ear by the introduction of names and properties not sufficiently familiar, and when at the same time it retains novelty enough to strike." She targets Aikin's assertion that through accurate observation poets could become better, more informed naturalists, than the naturalists themselves. This isn't to say that she thought poets incapable of surpassing naturalists' knowledge of nature. However, Barbauld urges the need for readers' "familiarity" with the natural objects poetically described. She explains that her concern arises from having "seen some rich descriptions of West Indian flowers and plants,—just, I dare say, but unpleasing merely because their names were uncouth, and forms not known generally enough to be put into verse."[18] By designating these plant names as "uncouth," Barbauld may allude to contemporary poets' aesthetic problem of deciding whether to employ species' scientific or common names in verse. If so, then she echoes a warning voiced in Aikin's *Essay,* when he favors a West Indian tree's Latin name, "Palma Maxima," rather than "wretchedly degrad[ing]" its "dignity and grandeur" with "its vulgar name of the Cabbage tree!"[19] This implies that poets should choose the most aesthetically appealing species name, whether Latin or common.

At the same time, "uncouth" may connote that which is unfamiliar, and since Barbauld specifically refers to West Indian plants, both the common and scientific names of these species would be unfamiliar to most British readers. Recommending avoidance of natural objects "not known generally enough to be put into verse," she makes a degree of familiarity requisite to this "novel" poetry and thereby qualifies the natural-historical knowledge available for use by the poet. When the poet's scientific knowledge, gained through observation, surpasses that of the naturalist, the poet is in danger of overreaching his or her bounds, that is, overreaching the poet's function for the reader. She cautions, "It is not, I own, much to the credit of poets,—but it is true,—that we do not seem disposed to take their word for any thing, and never willingly receive *information* from them."[20] For Barbauld, poetry is not an appropriate venue for new scientific findings. While natural history can bring originality to verse, incorporated scientific observations must not be so novel that they detract from the reader's pleasure in the poem by provoking confusion in unfamiliarity of facts or doubts about the subject's validity.

In this vein, Barbauld's "To Mrs. P[riestley], With Some Drawings of Birds and Insects" again presents a model, demonstrating how to avoid uncouth names and unfamiliar forms in descriptive nature-poetry. Her poem portrays several birds and insects without providing enough detail for readers easily to distinguish the exact kinds. Seeking to identify the species Barbauld intended, her most recent editors, William McCarthy and Elizabeth Kraft, cite Pennant's zoology to corroborate her bird depictions and suggest she "may have derived her West Indian insect descriptions from . . . the Dutch naturalist Maria Merians" (44). However, while their classifications, especially of the bird species, are very plausible, I find it more interesting that Barbauld's descriptions may be considered (to reprise her "Lesson in the Art of Distinguishing") "so loosely worded that we cannot certainly tell what animals are meant by them." She offers only general names for birds, such as Eagle and Pheasant, without denoting particular species. In Pennant's zoology, he lists three different species of eagles, so that the burden is on the reader to piece together clues from Barbauld's description if one wishes to decipher which eagle is "meant" by the verse. Consulting Pennant, McCarthy and Kraft hypothesize with convincing accuracy that Barbauld has the Golden Eagle in mind, for she places her eagle "On Snowden's rocks, or Orkney's wide domain," and "Pennant notes its presence in the Orkney Islands and occasional appearances 'in *Snowden* hills'" in Wales (l. 33).[21] More difficult to conjecture, however, are the identities of Barbauld's West Indian insects. She describes "the proud giant of the beetle race" with "shining arms" and "spreading horns," as "his rich treasury swells with hoarded grain" (ll. 113, 114, 117, 120). Yet, this characterization would seem to fit a number of beetle species in Britain as well as in the West Indies.

In her portrayals of both birds and insects, Barbauld provides just enough specificity to elicit familiarity. Her details of environment, food, physical characteristics, and behavior offer sufficient information that explanatory footnotes revealing the particular species she has in mind would seem appropriate, yet she supplies none. Presumably, Barbauld thought the species described either familiar enough to require no footnote (as perhaps in the case of the birds), or that her depiction presents "a lively picture of an object, as if you saw it" and need not distract with "uncouth" specificity (as she may have feared for her description of West Indian insects). Her poem displays accurate observation and knowledge of natural history, balancing imaginative pleasure with a paradoxically familiar novelty of information—that is, an ability to make the familiar seem novel through closeness of observation. In Barbauld's *Essay on Akenside's Poem on the Pleasures of Imagination* (1795), she argues that "the Muse would make a very indifferent school-mistress. Whoever therefore reads a Didactic Poem ought to come to it with a previous knowledge of his subject; and whoever writes one, ought to suppose such a knowledge in his readers."[22] For her, poetry should not be pedagogical or informative in any subject, let alone in technical science. Indeed, her essay distinguishes that "the end of Poetry is to please, and of Didactic precept the object is instruction."[23] Poets writing didactically risk alienating their audience through unfamiliar ideas. In Barbauld's own theory and practice, even as she adapts natural history to verse, her primary purpose in poetry is to *please,* whereas that of her scientific prose is to *teach.*

Natural Theology and Teaching Nature's Prose

Adhering to her critical distinction between these genres, Barbauld writes educational texts in prose rather than poetry, expressing concern for children's learning processes. Sixteen years after *Poems* (1773) brought Barbauld national renown, John Aikin declared countless parents' gratitude for her subsequent "condescension" to pen pedagogical works.[24] Many succeeding generations in Britain and America associated her name with vivid childhood memories of learning to read and to piously relate the self to the natural world through her *Lessons for Children* (1778) and especially her *Hymns in Prose for Children* (1781). By writing in prose, Barbauld claimed to correct an error in what she deemed the only other book "calculated to assist" children in forming religious consciousness, Isaac Watts's *Divine Songs attempted in easy Language for the Use of Children* (1715), which conveys devotionals in poetry. She composes *Hymns* in prose rather than verse, explaining, "the very essence of poetry is an elevation in thought and style" that should not "be lowered to the capacities of children," nor should children read poetry until capable of appreciating it (237). Barbauld thus educates in prose in order to preserve poetry's integrity, as well as to suit the child's developing mind. Her insistence that readers "never willingly receive *information*" from verse also doubtless influenced this pedagogical decision. Through prose, Barbauld explores themes that shaped her own youth and methods of teaching.

The natural sciences played a prevalent role in Barbauld's early life and education. She was born into a family of Presbyterian Dissenters, and when she was fifteen her father, the Reverend John Aikin, became tutor in languages and belles letters, and subsequently in divinity, at the Dissenting academy at Warrington, Lancashire. The Warrington school boasted prowess in various scientific fields. Corresponding with the English botanist Richard Pulteney, Barbauld's father discussed herbariums, including Linnaeus's *Systema Naturae* (1735), *Genera Plantarum* (1737), *Species Plantarum* (1753), and *Amoenitates* (1763), and he objected that Linnaeus's theory of the creation of plants and animals was incompatible with the Mosaic account.[25] His botanical pursuits and eager engagement with science's religious implications encouraged his children's interest in natural history that permeates their later literary endeavors. And of course it was at Warrington that Barbauld became friends with Joseph Priestley, who tutored there from 1761 to 1767, while performing experiments in natural philosophy. Exposure to science influenced Barbauld's teaching in turn, and when she and her husband later ran a boys' school at Palgrave, she incorporated the natural sciences, enlivening geography with detailed descriptions of "the natural history of animals."[26] While at Palgrave, she also wrote *Hymns in Prose,* synthesizing her dissenting beliefs and this classroom instruction.

In *Hymns,* Barbauld seeks to teach the child reader "to see the Creator in the visible appearance of all around him, to feel his continual presence, and lean upon his daily protection" (237). By conditioning children to appreciate God through his works in nature, Barbauld employs a brand of natural theology we may connect with the late-seventeenth-century British naturalist John Ray. Ray, like Barbauld, had been a Dissenter who viewed science as "a means to the worship of God."[27] Pennant lauds him as the founder of British zoology, and Ray also notably advanced the knowledge and classification of fish, insects, and plants, and has been recognized as the father of British natural theology for his enduringly popular text, *The Wisdom of God manifested in the works of Creation* (1691).

Much of Barbauld's *Hymns* for children reads as a simplified version of the natural-theological tenets of Ray's *Wisdom*.[28] When Ray, for instance, references Psalm 148, which calls upon the sun, moon, stars, mountains, trees, and all creatures "to praise the Lord," he asks the rhetorical question, "How can that be? Can senseless and inanimate things praise God?"[29] Ray resolves this biblical conundrum by explaining that these creations afford "matter or subject of praising [God], to rational and intelligent beings."[30] If nature cannot praise, it incites humans to praise the Creator of such works. Just so, Barbauld's *Hymns* instructs children that "the birds can warble, and the young lambs can bleat; but we can open our lips in his praise, we can speak of all his goodness. Therefore we will thank him for ourselves, and we will thank him for those that cannot speak" (240). In Ray's text, he goes on to state that "man is commanded to consider [vegetables, beasts, birds, and insects] particularly, to observe and take notice of their curious structure, ends, and uses, and give God the praise of his wisdom, and other attributes therein manifested."[31] Addressing the "child of reason," Barbauld likewise advocates close attention to nature, not merely for the sake of gaining knowledge of nature, but also to attain knowledge of God. When the child recounts the various plants and animals spotted during his walk through the meadow, the educating narrator chides, "Didst thou observe nothing besides? Return again, child of reason, for there are greater things than these.—God was among the fields; and didst thou not perceive him? . . . God is in every place; he speaks in every sound we hear; he is seen in all that our eyes behold: nothing, O child of reason, is without God" (245, 246). Emphasizing knowledge of God through nature, Barbauld's prose also imparts specific scientific details.

In Hymn IX, Barbauld describes an imaginative excursion, elucidating the physical structure and environmental location of various trees and plants, using their common names, such as fir, grey willow, mallow, iris, water-lilies, and so on. Organizing the snowdrop, primrose, carnation, and laurustinus according to a calendar of flora, she marks the chronology of plants' development, explaining, "They are marshaled in order: each one knoweth his place, and standeth up in his own rank" (251). Her taxonomic rhetoric of "order" and "rank" alludes to more advanced botanical lessons like those in her essay "On Plants," and arguably exposes young readers to conservative conceptions of social order in which "each one knoweth his place." She paradoxically promotes both the (religious) impossibility of fully understanding God's works, and the (Linnaean) quest to "discover" all the world's species and configure an order reflective of divine design: "They that know the most, will praise God the best; but which of us can number half his works?" (252). Barbauld's question to the child reader implies that even if the taxonomic task of "number[ing] half his works" remains always out of reach, the tacit point, as exemplified by her own pedagogical prose lists and descriptions of plants, is to try.

"THY BOUNDED SPHERE": WOMEN'S EDUCATION, WOMEN'S POETRY

Barbauld's *Hymns* does not target one sex over the other, but both her poetry and prose demonstrate great awareness of society's gendered divisions in scientific education. In her own studies of natural history at the Warrington academy, she was not permitted to enroll in this all-male institution but frequently conversed with students, tutors, and male family members and friends, and she later recommended such discussions as integral to women's intellectual development.[32] Barbauld took advantage of this academic environment, reading broadly, teaching herself Latin and Greek, and absorbing the school's strength in the natural sciences.[33] However, her poetry sometimes displays a subtle tension in her relation to science due to her sex.

Several scholars note Barbauld's influence over her younger brother, John Aikin, as well as her possible jealousy of his greater educational and professional opportunities. In her poem, "To Dr. Aikin on His Complaining that She Neglected Him, October 20th, 1768." Barbauld barely smothers her complaint against the siblings' inequality due to sexual, not intellectual, difference:

> Our path divides—to thee fair fate assign'd
> The nobler labours of a manly mind:
> While mine, more humble works, and lower cares,
> Less shining toils, and meaner praises shares.
> Yet sure in different moulds they were not cast
> Nor stampt with separate sentiments and taste.
> But hush my heart! nor strive to soar too high,
> Nor for the tree of knowledge vainly sigh;
> Check the fond love of science and of fame,
> A bright, but ah! a too devouring flame.
> Content remain within thy bounded sphere,
> For fancy blooms, the virtues flourish there. (ll. 50–61)

Barbauld consoles herself by suggesting, as she does elsewhere, that her "bounded sphere" nevertheless particularly suits her to the occupation of poetry.

Combating women's educational inequality in the natural sciences, Barbauld sometimes couches scientific assertions in humor or satire, a strategy that mitigates the female poet's stance of intellectual superiority. For example, her poem in celebration of Warrington, "The Invitation" (1773), represents the poet as the ultimate naturalist, scrutinizing the school's scientific students as natural objects for her own study: "Some pensive creep along the shelly shore;/Unfold the silky texture of a flower;/With sharpen'd eyes inspect an hornets sting,/And all the wonders of an insect's wing" (ll. 155–58). In observing observers of nature Barbauld depicts her place outside of the academy, of the sanctioned attainment of this scientific knowledge and, despite this exclusion, assumes a privileged position over these (male) naturalists who creature-like "creep" as she records their movements in her natural description. She adopts a female perspective of poetic distance that analyzes, and converts to entertainment, both the science and the scientists themselves.[34]

Barbauld argued that it was imperative for young women to be educated in "natural history, astronomy, botany, experimental philosophy, chemistry, physics" (480).[35] Indeed, in the last quarter of the eighteenth century, "botany was considered a science particularly suited to women" and was prescribed for women to "provide pleasure and instill virtue."[36] Of course this association of women, botany, and pleasure famously provoked criticism from conservatives such as Richard Polwhele, who, in his poem *The Unsex'd Females* (1798) exasperatedly wonders "how the study of the sexual system of plants can accord with female modesty." Linnaeus's "sexual system" of botany taxonomized plants according to their reproductive parts, determining a flower's class based on the number and proportion of (male) stamens, and its order by counting the (female) pistils. The simplicity of this method of classification, combined with Linnaeus's personification of plants' sexual relations through marriage metaphors in which, for example, male and female organs became "husbands" and "wives," contributed to the system's popularity.[37]

Barbauld's separate listings of natural history and botany indicate her endorsements both of botany's suitability for women as well as that of zoology, including the branches of ornithology and entomology. Assuaging fears of impropriety, Barbauld's letters "On Female Stud-

TO TEACH AND TO PLEASE + 35

ies" argue that women who learn about natural history specifically will "take what belongs to sentiment and utility" and "feel the mind struck with lively gratitude," observing God through his creation. She informs young women that natural history "will give you an interest in every thing you see. If you are feeding your poultry, or tending your bees, or extracting the juice of herbs, with an intelligent mind you are gaining real knowledge; it will open to you an inexhaustible fund of wonder and delight, and effectually prevent you from depending for your entertainment on the poor novelties of fashion and expense" (480). Insisting on the primacy of "sentiment and morals," Barbauld justifies natural history as a decorous female pursuit that incites productivity as well as virtuous, intellectual pleasure, counteracting the corruptive attractions of fashionable society.

By emphasizing sensibility and domestic knowledge, Barbauld upholds sexual divisions of labor, asserting the merit of women's industry and extending its scope to literature. She delineates women's education, advertising the separate studies of a young man and a young woman "to be chiefly fixed by this,—that a woman is excused from all professional knowledge." As her most recent editors suggest, Barbauld's dry statement of women's preclusion from professions of science, law, politics, and so forth in contemporary British society is a pronouncement of fact rather than an endorsement of this status for women (474). Operating within these prescribed bounds, Barbauld elucidates women's potential to excel. Although men may claim specialized professional knowledge, "which is nowise valuable in itself, but as a means to that particular profession," women arguably have the upper hand, for "a woman ought to have that general knowledge of [all studies] which marks the cultivated mind" (481). Moreover, according to Barbauld, a woman's "cultivated mind" "fit[s] her in a peculiar manner for the worlds of fancy and sentiment, and dispose[s] her to the quickest relish of what is pathetic, sublime, or tender." To women, "therefore, the beauties of poetry, of moral painting, and all in general that is comprised under the term of polite literature, lie particularly open" (477). By emphasizing women's superior sensitivity in all ranges of feeling, especially in sentiment's association with imagination and morality, Barbauld asserts a claim to women's "profession" in poetry.

For Barbauld, the necessity of pleasure to poetry shapes science's function in verse, and helps appropriate poetry as the territory of women. To illustrate this further, I would like to offer a new way of understanding what has become perhaps Barbauld's most infamous poem, "To a Lady, With Some Painted Flowers" (1773). Nearly twenty years after its publication, Mary Wollstonecraft attacked Barbauld's poem, and quoted it in full in the notes to her *Vindication of the Rights of Woman* (1792):

> Flowers to the fair: To you these flowers I bring,
> And strive to greet you with an earlier spring.
> Flowers sweet, and gay, and delicate like you;
> Emblems of innocence, and beauty too.
> With flowers the Graces bind their yellow hair,
> And flowery wreaths consenting lovers wear.
> Flowers, the sole luxury which nature knew,
> In Eden's pure and guiltless garden grew.
> To loftier forms are rougher tasks assign'd;
> The sheltering oak resists the stormy wind,
> The tougher yew repels invading foes,
> And the tall pine for future navies grows;
> But this soft family, to cares unknown,
> Were born for pleasure and delight alone.
> Gay without toil, and lovely without art,
> They spring to cheer the sense, and glad the heart.
> Nor blush, my fair, to own you copy these;
> Your best, your sweetest empire is—to please.

Wollstonecraft denounced Barbauld's botanical analogy as an "error . . . which robs the whole [female] sex of its dignity, and classes the brown and fair with the smiling flowers that only adorn the land. This has ever been the language of men, and the fear of departing from a supposed sexual character, has made even women of superior sense [Barbauld] adopt the same sentiments."[38] According to Wollstonecraft, Barbauld presents women as specimens to be objectified and classified alongside plants and other natural objects in "language" that dehumanizes by differentiating and subjugating the female sex.

However, an alternative reading of the poem, that Wollstonecraft seems to miss, more readily aligns with Barbauld's theories about poetry and about the separate educational goals for young men and women by applying her botanical analogy as much to poetry itself as to women. The "Painted Flowers" of the poem's title, brought to an unknown "Lady," may refer to additional verses on the subject of plants (as in Barbauld's "To Mrs. ——, on Returning a fine Hyacinth Plant after the Bloom was Over") through *ut pictura poesis*. For families that could afford such tutelage, the drawing and painting of flowers had become a fashionable part of young women's education.[39] Barbauld alludes to this feminine practice of painting, underscoring her verses' adoption of natural history's descriptive power "to give you a lively picture of an object." The poem's meaning thus easily expands, comparing her painted "flowers" with poetry as well as with women. Through this association, poetry's purpose (as well as women's) is to provide "pleasure and delight," anticipating Keats's insistence on verses' ease of inspiration, as Barbauld claims that poetry should be "Gay without toil, and lovely without art" (ll. 14, 15).[40] If she endorses an existing, sexist equation of women with nature, she also arguably shifts the analogy's power structure to make nature-poetry the province of women writers. Since, for Barbauld, the primary purpose of poetry, and of nature-poetry in particular, is to produce pleasure, who better to fulfill this literary task than women, whose very "empire," she states, in a ventriloquization that converts degradation into advantage, "is—to please" (l. 18). Dichotomizing body and mind, she confines men to merely physical exertions, a category conventionally imposed on women, so that while men are "rougher tasks assign'd" of defending the nation from "invading foes," women are here, as in "On Female Studies," afforded "more leisure" to pursue knowledge and literary accomplishment (ll. 9–11). An ironic undertone persists in Barbauld's poetic allusions to men's "loftier," professional endeavors. Rather than calling for the dismantling of sexual hierarchies with the overt gusto that exposed Wollstonecraft to immediate censure, Barbauld urged that women's "empire" is "Felt, not defined, and if debated, lost," working within traditional distinctions among the sexes to procure less obvious, but more functional, leverage.[41] Barbauld acknowledges her own motivational role of providing, through these flowers, poetic models for women to emulate in the "profession" of poetry, stating in her penultimate line, "Nor blush, my fair, to own you copy these" (l. 17). As William McCarthy explains, numerous female poets wrote enthusiastic verse tributes to Barbauld and credited her with inspiring their poetry.[42]

Barbauld's subtle reformulations of women's literary opportunities within this framework of poetic pleasure and natural history may have escaped Wollstonecraft's detection, but Polwhele condemned her alongside Wollstonecraft in *The Unsex'd Females*. Barbauld confounded others, like the reviewer of her *Poems* (1773), who complained of not being able

A flower painting (1759) by Mary Moser, perhaps the most famous British female painter of her day, particularly known for her floral depictions, and one of only two women included as founding members of the Royal Academy of Arts. (RSA, London)

to find the *"Woman"* in her verse.[43] Her gender came under further attack when Coleridge, Lamb, and Southey mistakenly attributed to her a negative review of Lamb's *John Woodvil* (1802), and subsequently perverted her name to "Bare and Bald," attempting to neutralize her literary and critical power through an assault on her femininity.[44] These men's efforts to de-sex Barbauld demonstrate the perceived threat of her literary influence and of her expanded territorial claims for women in science and literature.

THE SCIENCE WHICH IS NOT ONE

Barbauld's employment of science in literature anticipates later efforts to make both fields more accessible to women. Her *Poems,* published in 1773, predate by sixteen years Erasmus Darwin's long poem *The Botanic Garden* (1789), which inspired many women's verses on natural history. As other scholars note, Darwin's poetic depictions of flowers as human analogues in *The Loves of the Plants,* the second part of *The Botanic Garden,* reinforce established female stereotypes, not in themselves particularly liberating.[45] However, Darwin directly appeals to women, describing his work as "diverse little pictures suspended over the chimney of a Lady's dressing-room," and thus his goal "to induce the ingenious to cultivate the knowledge of BOTANY" struck many women as sanctioning their poetic and intellectual participation in this scientific field.[46] As I further discuss in chapter 4, numerous natural history poems preceded Darwin's; nevertheless, women such as Arabella Rowden, Sarah Hoare, and Charlotte Smith wrote verses specifically responding to his popular structure and theme. In her published letters, Anna Seward declared that Darwin's poem fulfilled "the union of natural history and of modern philosophic science with poetry" as set forth in Aikin's *Essay on the Application of Natural History to Poetry.*[47]

Barbauld herself was "quite fascinated" with Darwin's versification of the Linnaean sexual system of plants and "talked of it with rapture," for which she was, in fact, "scolded" by Samuel Rogers.[48] It is, however, unsurprising that Barbauld found Darwin's poem enticing. Whereas Aikin, like Pennant, favored the versification of zoology as "the noblest part of natural history" and discouraged botanical poetry, Barbauld rejoined, "I should not have confined the track quite so much as you have done to the animal creation, because sooner exhausted than the vegetable; and some of the lines you have quoted from Thompson [*sic*] show with how much advantage the latter may be made the subject of rich description."[49] Darwin's poem, versifying a naturalist's system already familiar to the public and focusing on botany in particular, thus fulfills Barbauld's vision for natural history poetry more than Aikin's. Moreover, her approval makes sense in terms of the poem's formal arrangement, which upholds her ideological division between poetry and prose.

In *The Botanic Garden,* Darwin elaborates his imaginative verse with scientific prose footnotes, creating a structural division that correlates with Barbauld's distinction between pedagogical prose and pleasurable poetry. In his poem's first "Interlude," Darwin generates a dialogue between two characters, the "Bookseller" and the "Poet," in which the Bookseller asks, "what is the essential difference between Poetry and Prose?"[50] Darwin's answer to this question recalls Barbauld's differentiation of description from definition in her "Lesson in the Art of Distinguishing" as well as her early concerns about the extent to which science should enter verse. Darwin asserts that "the Poet writes principally to the eye, the Prose-writer uses more abstracted terms. . . . Science is best delivered in Prose, as its mode of reasoning is from stricter analogies than metaphors or similes."[51] Metamorphizing plants into humans, Darwin puts this distinction into practice. His poetic plants are so divorced from the content of his informative prose footnotes that one modern scholar, looking for botany in Darwin's verse, felt compelled to ask, "where have all the flowers gone?"[52] Darwin's separation of poetry and prose enacts Barbauld's theoretical distinction when she writes that poetry of natural description "is intended to give you a lively picture of an object," while prose scientific definitions adhere to abstract characterizations, lending themselves to classification and pedagogy. Significantly, the divergence of scientific abstraction and visual particularity also formed the center of a philosophical debate popularized over a century earlier, and that continued to provoke naturalists' argument into the first quarter of the nineteenth century.

In this controversy's earlier manifestation, John Locke gained the enmity of classifiers ordering groups in nature according to a single, "essential" character to create artificial taxonomies when he rejected the universal notion of species in his *Essay concerning Human Understanding* (1689).[53] For him, "species" constitutes a useful category in abstract philosophical discussion, but does not exist in nature. The problem, according to Locke, with an abstract, taxonomic conception of species is the incommensurability between the Particular/Individual (Tom,

Dick, and Harry) and the Universal (Man). His objections pose a fundamental problem to the attainment of knowledge, for if we can only directly know individuals, and our experiences cannot be generalized by converting particulars into universals, then science ceases to exist. Locke's ideas inspired John Ray to reconceive his notion of species and, in turn, influenced Thomas Pennant and Georges-Louis Leclerc, comte de Buffon, among other important naturalists of the eighteenth and early nineteenth centuries. Buffon followed Ray in rethinking species, incorporating as many particulars as possible before moving to a universal, and established something more akin to a network than a hierarchical taxonomy for configuring relations between species.[54] I further explore the ideas of Ray, Pennant, and Buffon, and their impact on women writers in subsequent chapters. However, most interesting to the present argument is that Barbauld and many other women authors also closely attend to various particulars and individualizations prior to panning out to universals. To state that male Romantics such as Wordsworth often move more quickly to abstractions while women poets tend to spend more time lingering on particularities repeats Stuart Curran's classic argument in his essay "The I Altered."[55] Yet, the theory becomes refreshed and refined here by connecting literary abstractions with those of science, and by noting that, in her literary criticism, Barbauld perceived poetic abstractions as "scientific" in a way that counters her own efforts in natural history poetry.

Barbauld's "Prefatory Essay" to *The Poetical Works of Mr. William Collins* (1797) offers valuable insights into her understanding of particulars and abstractions in poetry, as well as their function in science. She blends knowledge of natural history in this essay of literary criticism, identifying hothouse flowers that "grow wild in many parts of Persia," and correcting Collins's ornithological terminology.[56] Exploiting overlaps in vocabularies of scientific taxonomy and literary criticism (a subject further explored in chapter 3), she argues, "The different species of Poetry may be reduced under two comprehensive classes."[57] Within this poetic taxonomy, Barbauld's first class of verse includes didactic, dramatic, epic, moral, and descriptive poetry, that is, "descriptions of natural objects, where the mind recognizes with pleasure the forms and colouring it admires in the various scenes and productions of the visible world."[58] In contrast, she determines that the second class "consists of what may be called pure Poetry, or Poetry in the abstract," a category incorporating "All that is properly *Lyric Poetry*."[59] It is in this second class that Barbauld places the poetry of Collins, who himself occupies "a respectable rank amongst our minor Poets" and receives her controlled praise.[60]

Although Barbauld designates "Poetry in the abstract" as "pure Poetry," she undermines this term of seeming predilection. Unlike the descriptive moral and nature-poetry of the first class, the so-called pure poetry is "obscure," of a "shadowy nature," "conversant with an imaginary world, peopled with beings of its own creation" based in the poet's subjective sentiments, and thus inaccessible to readers who do not share "similar contemplations."[61] According to Barbauld, lyric poetry will never be popular because even "the most beautiful *Ode* will only please those who by being long conversant with the best models of Poetry in a polished age, have acquired a scientific and perhaps, in some degree, a factitious taste."[62] It is important to notice Barbauld's placement of pleasure, and what she considers "scientific" in this critique. Whereas descriptive poetry enlivens the mind with pleasurable recognition of the natural world, sentiments with which all readers may empathize, poetry of the abstract can only please those who approach poetry scientifically and artificially. For Barbauld, lyric poetry enters too deeply into the abstract character of science or knowledge, and when abstractions get in the way of pleasure poetry is compromised.

Barbauld's caution to Aikin about science's engagement with verse, and the need for scientific information to be familiar to readers, seeks a balance between the particular and the abstract, between accurate observation and imaginative pleasure, portraying the paradox of familiar novelty in nature. In this balance, Barbauld avoids a trap that Wollstonecraft feared had been set for women when the female sex was sometimes touted as skilled in perceiving particulars and thus suited for accurate observations in the natural sciences.[63] Worrying that such assertions may present women as incapable of conceiving in abstract terms, Wollstonecraft writes that "the power of generalizing ideas, of drawing comprehensive conclusions from individual observations . . . has not only been denied to women; but writers have insisted that it is inconsistent, with a few exceptions, with their sexual character."[64]

In Barbauld's poetry, the question is not whether she deals in abstractions but what place and degree of emphasis such generalizations assume, and this can be exemplified in her affecting poem "The Caterpillar."[65] Here, her "sharpened eye," having been employed in "persecut[ion]" and "slaughter" of caterpillar "tribes and embryo nations," softens to a feeling curiosity when examining the particular physical qualities of a single individual:

> No, helpless thing, I cannot harm thee now;
> Depart in peace, thy little life is safe,
> For I have scanned thy form with curious eye,
> Noted the silver line that streaks thy back,
> The azure and the orange that divide
> Thy velvet sides; thee, houseless wanderer,
> My garment has enfolded, and my arm
> Felt the light pressure of thy hairy feet;
> Thou hast curled round my finger; from its tip,
> Precipitous descent! with stretched out neck,
> Bending thy head in airy vacancy,
> This way and that, inquiring, thou hast seemed
> To ask protection; now, I cannot kill thee. . . .
> A single wretch, escaped the general doom,
> Making me feel and clearly recognise
> Thine individual existence, life,
> And fellowship of sense with all that breathes. (ll. 1–13, 24–27)

In a way that arguably mimics the creation of scientific laws, Barbauld moves from the particular to the abstract and universal. Through the personal contact of experiencing the "light pressure of [its] hairy feet," examining the coloration of its "velvet sides," and observing its ability to seem, almost humanly, "inquiring" and pleading for protection, the caterpillar becomes real—a living creature outside the self, worthy of care and recognition. It attains an identity separate from the nameless masses and from the poet's consciousness, while simultaneously becoming a means to the poet's expanded knowledge of herself and "fellowship" with the world around her. Barbauld's reader comes away with a vivid, "very full," "lively picture of [the] object, as if you saw it," and a pleasing sense of familiarity that imparts the feeling of having been "touched" by this "single wretch." As in Barbauld's depictions of birds and insects in her poem, "To Mrs. P[riestley]," her vivid description of the caterpillar indicates that she has a distinct species in mind, but she again omits such specific information, balancing particulars with generalities. Her final lines broaden to a larger moral framework so that minute interactions with this caterpillar elicit compassion for each "individual existence, life." The verses thus create novelty through careful union of moral sentiment and attentiveness to nature, of universals and particulars that pervade Barbauld's poetic oeuvre and that she used to justify women's verse and attainment of scientific knowledge in her letters "On Female Studies."

"WHERE SCIENCE SMILES, THE MUSES JOIN THE TRAIN": A QUALIFIED APPROACH

Despite her acumen in natural history, Barbauld urged caution in combining science and poetry, refusing to identify the poet too closely with the naturalist. Her descriptive nature-poetry subtly recalibrated women's relations to nature and pleasure, and influenced subsequent female writers who looked on her precedent as one to be followed, repudiated, and modified for the next four decades in what Judith Pascoe has recognized as a "literary movement . . . of British women's writing merging poetry and science."[66] Of equal interest is Barbauld's influence on the period's male poets.

While Lamb confines Barbauld to a rational, scientific mode of stifled imagination, Wordsworth arguably presents a more complex case of comparison. In the preface to the second edition of *Lyrical Ballads* (1800), Wordsworth famously defines the poet as "a man speaking to men," and writes that poetry's only "necessity" is "producing immediate pleasure." In this way, he seeks to masculinize both poetry and pleasure, especially an imaginative form of pleasure that cannot be separated from expressions of knowledge and sympathy. As Geoffrey Hartman notes of Wordsworth, "he rarely counts the streaks of the tulip, but he constantly details the state of his mind."[67] Like the observant poet-naturalist Aikin advocated as embodied by Barbauld, Wordsworth states, "I have at all times endeavored to look steadily at my subject," but he clarifies that his "subject" is "the manner in which our feelings and ideas are associated in a state of excitement," thereby emphasizing the predominance of internalization.

Wordsworth follows Barbauld in exploring the division between the Poet and the "Man of Science" (Wordsworth's phrase), and in voicing concern about unfamiliar scientific information in poetry. However, while Barbauld indicates that some natural-historical subjects *are* familiar enough for verse, affirming in her poem "The Invitation" that "Where science smiles, the Muses join the train" (l. 109), this is an admission that Wordsworth denies in his preface. He employs a series of poignant "if" clauses that endlessly defer the mixing of poetry and science, stating, "If the labours of Men of Science ['the Chemist, the Botanist, or Mineralogist'] should ever create any material resolution. . . . if the time should ever come when what is now called science, thus familiarized to men, shall be ready to put on, as it were, a form of flesh and blood, the Poet will lend his divine spirit to aid the transfiguration."[68] For Wordsworth, as these "if" clauses make clear, science is not ready to appear in verse; rather, according to him, it is yet too solitary, separate, and abstruse to become "flesh and blood."

Scholars have attributed Wordsworth's theoretical opposition to scientific poetry in the preface to his jealousy of Sir Humphry Davy, or to his distaste for the "gaudiness" of language in Erasmus Darwin's poems.[69] While both of these premises are convincing, Wordsworth's statements also arguably work to de-legitimize women's versifications of natural history, consciously reclaiming poetry as a more strictly imaginative, masculine vocation. His "if" contingencies condemning scientific subjects as improper, unfamiliar, and premature for poetry ignore the fact that by 1800 a number of writers (many of them women) already had successfully and popularly committed such scientific themes to verse. Wordsworth represents science's developing professionalization around the turn of the nineteenth century as producing esoteric knowledge, inaccessible to poets, and particularly to women, because such subjects are "cherishe[d] and love[d]" by the "Man of science" in isolation, as opposed to the more communal efforts of poetry. Unlike Barbauld, he does not allude to women's preclusions from science in order to critique educational inequality. Wordsworth instead suggests a perversion of affections in male scientific professionals because they "cherish and love," not women or even humanity but solitary pursuits. He negatively characterizes this isolation in contrast with his more positive portrayal of the sociable (male) poet. Paradoxically, Wordsworth seems accurately to assess the difficulties of versifying scientific subjects once they have retreated within the academies. After all, the professionalization of science in the early nineteenth century historically inhibited the participation of amateur naturalists that had been so important to Aikin's previous call to verse and to women's inclusion in the study and discovery of scientific phenomena. Without acknowledging science's sociable past, Wordsworth seizes on exclusionary practices of science's professionalization that would undermine the serious pursuit of scientific poetry.

Barbauld anticipates Wordsworth in critiquing science's tendency toward abstraction and potential for insularity and exclusion, as well as in voicing the need for its familiarity in the public imagination before entering verse. She, like him, in his preface's theory of poetic diction where he calls for "language near to the language of men," disapproved of the "uncouth," whether in names or obscurity, as a distraction from poetic pleasure. Yet, for Barbauld, the primacy of pleasure in poetry qualifies, not precludes, science's role in verse. She thus cautiously facilitates the beginning of this era's movement of women merging literature and natural history. Barbauld's moderate views arguably function as a middle point, a comparative touchstone between those of male Romantics such as Lamb and Wordsworth, on one hand, and those of women writers such as Maria Riddell and Charlotte Smith, on the other. Smith, for instance, frequently includes species' scientific names in her verse and blurs Barbauld's line between pleasurable poetry and pedagogical prose. Such differences help chart the spectrum of women's aspirations for literary originality through natural history and their unique relations to traditional narratives of Romanticism. Displaying science's growing possibilities for women's literature, the next chapter makes an excursion away from Britain to the West Indies, showing how Maria Riddell's poetry and travel writing contributed to scientific notions of gender, race, and nation. Whereas Barbauld exemplified West Indian species as unfamiliar and therefore inappropriate for verse, and worked to distinguish poetry and prose, the West Indian islands' national and biological hybridity inspired Riddell to engage with hybridity's potential originality in both scientific theory and literary form.

2

Hybrid Britons

West Indian Colonial Identity and Georgic
Originality in Maria Riddell's Natural History

While travel literature typically describes new geographies, biological species, and scientific discoveries in its accounts of foreign nations, and was recommended by John Aikin and other critics to be read by poets seeking original subjects for verse, Samuel Johnson portrayed travel writing of the late eighteenth century as another genre increasingly devoid of novelty.[1] Johnson thus exhorted patriotic travelers to "mingle pleasure with instruction" and report differences separating Britain from other countries in aspects of "human life," including "works of genius," agriculture, medicine, manufacture, customs, and policy, as a means to novelty, knowledge, and national improvement. In the previous chapter, I discussed Anna Barbauld's caution that such detailed explorations of unfamiliar subjects are suited for prose and not verse forms. However, other poets embraced these distant countries' exoticism as a clear opportunity for originality and for advocating national and scientific advancements in the tradition of georgic poetry and its formal hybridity that often mixed with the pastoral and included explanatory footnotes or endnotes.[2] Formal hybridity increasingly became a point of contention in both literature and natural history. At this time, most naturalists believed in the fixity of biological species, that is, in the constancy and unalterability of species. Yet some naturalists also began to explain species' differences through possibilities of change and mixture. In her *Voyages to the Madeira, and Leeward Caribbean Isles: with Sketches of the Natural History of these Islands* (1792), Maria Riddell combines familiar and unfamiliar information, landscapes, and ideals, employing concepts of biological hybridity to attain alternative means to literary originality and a politico-scientific framework for reconceiving both British nationalism and the often-denigrated Caribbean colonies.[3]

Indeed, death, disease, and degeneration permeate late eighteenth-century portrayals of the British West Indies and, as several scholars have recently suggested, these threats incited anxiety concerning hybridizing effects on relocated British subjects.[4] Arguably more alarming than the risk of physical alterations wrought by disease was that of degeneration in British national character, resulting in indolence, excessive passion, and barbarous corruptions related to a slave society. In natural history and medical tracts, such national degenerations found comparison with natural degenerations of zoological species. Theorizations of biological mutability were largely influenced by the work of Georges-Louis Leclerc, comte de Buffon, who famously associated the entire western hemisphere with biogeographical degeneration, an association that held significant implications for European colonists of these territories. Still, British writers sympathetic to West Indian interests often sought to reformulate the Caribbean capacity for alteration. Writers such as Edward Long, Janet Schaw, Bryan Edwards, and Riddell gesture toward conceiving of the West Indies as a space in which nationality acquires new meaning, where transplants of Scottish, Welsh, Irish, and English origin become agents of empire and "Britons" in a now-shared experience of common values and a broader sense of national (rather than regional) allegiances: a developmental shift of national identity that seems at home in these islands, where biological forms evoke the hybridity synonymous with island ecologies.

In *Voyages,* Riddell employs her knowledge of science to redirect contemporary conversations about hybridity. Through naturalists' zoological systems, she imposes a British nationalist, unifying agenda on the West Indian colonies and indicates the potential of these islands to foster the inherently hybrid British national identity. This interaction of nature and nation shapes Riddell's scientific attitude toward change in literary, racial, and biological forms. Subtly assuaging her British readers' fears of West Indian hybridity, she explores the improvements possible in British Caribbean islands through national values often associated with the georgic mode, such as Protestant industry, and projects the degenerations imputed to West Indians instead onto foreign (especially Catholic) nations. Within her travel narrative, the georgic becomes a political tool, not only modeling formal novelty and hybridity, but also allowing her to align different nations with classical, mock, and English georgics, designating relative cultural ideals and possibilities for improvement. Riddell interrogates even as she enforces the concepts of degeneration and improvement, questioning ideologies intrinsic to these formulations and their relevance to British colonies in the Atlantic. While some literary and scientific authors drew sharp distinctions between hybridity and originality, she acknowledges overlaps between these ideas. At the same time, Riddell's *Voyages* also includes noticeable elisions, particularly in regard to slavery, that complicate the presentation of these islands as conducive to the idealization of a unified Britain, and undercut the notion that the colonial periphery's nature and literature may serve as models for the metropole.

RIDDELL'S PERSONAL HYBRIDITIES

The poet and naturalist Maria Riddell (née Woodley; 1772–1808) embodied national hybridity.[5] Although she was born and educated in England, her family had long-established connections to the West Indies, and particularly to the Leeward Islands. Riddell's father inherited a plantation on St. Kitts and twice served as British governor of the Leeward Islands. Her mother may have been born on St. Kitts, as was her cousin Sir Ralph Payne, Baron Lavington, an Irish peer, who likewise was twice governor of these islands and owned a plantation on Antigua. In 1790, at nearly eighteen, Riddell was married on St. Kitts to a Scottish lieutenant and Antiguan plantation owner, Walter Riddell, and the couple moved to Scotland shortly thereafter. She began writing poetry early in life, and at only sixteen articulated a sense of hybrid nationality in her poem "Inscription Written on an Hermitage in the West Indies."[6] Longing for "Albion," with the term's mixed references to British regions and cultural histories, Riddell nevertheless "prefer[s]" the West Indian hermitage.[7] Her verses thus mingle affectionate ties to both Britain and its Caribbean colonies. Riddell's scientific travel literature also displays these dual allegiances, and her writing found encouragement when she and her husband relocated to Scotland, where she became the neighbor and close friend of Robert Burns.

Currently, in literary scholarship, Riddell is best known for her relationship with, and subsequent memoir about, Scotland's national bard. In his lifetime, Burns penned numerous poems for Riddell, and scholars long debated whether she was the recipient of his famous 'letter from

Portrait of Maria Banks Woodley Riddell (1806), by Sir Thomas Lawrence. (©National Trust/Angelo Hornak)

Hell" during their temporary rift following the "Rape of the Sabines" incident, in which Burns drank too much at a dinner party and offended the Riddell family.[8] This separation affected Burns deeply and, as his recent biographer states, "he was more than a little in love with her."[9] He praised Riddell's poetic talent and revered her as a friend and confidante. Additionally, and more importantly for my interests, Burns, who nearly sailed for the Caribbean in 1786 before hearing of the success of his first volume of poetry, provided the means by which Riddell's *Voyages to the Madeira, and Leeward Caribbean Isles: with Sketches of the Natural History of these Islands* came to be published. He introduced Riddell to the publisher of his first volume of poems, William Smellie, who also was a naturalist translating the works of Buffon and authoring *The Philosophy of Natural History* (1790). Immediately impressed with both the author and her work, Smellie wrote to Riddell, "If I had not previously had the pleasure of your conversation, the devil himself could not have frightened me into the belief that a female human creature could . . . have produced a performance so much out of the line of your ladies works. . . . [S]cience, minute observation, accurate description, and excellent composition are qualities seldom to be met with in the female world."[10] Although this line of praise justifies Hannah More's complaint that a woman writer's "highest exertions will probably be received with the qualified approbation, *that it is really extraordinary for a woman,*" Smellie's enthusiasm for Riddell's manuscript led him to become her friend and correspondent as well as her publisher.[11]

While Riddell originally intended to confine *Voyages* to a small circulation among her family and friends, Smellie persuaded her that the work deserved a larger audience and arranged for printing in Edinburgh and London. Upon her insistence, Smellie reluctantly contracted her name on the title page to "Maria R—," assuring her, "I still think that it would do great honour to any—in Britain."[12] This gesture toward anonymity did not keep critics from discovering Riddell's identity and pronouncing the work to be "not destitute of amusement," and "her proficiency in natural history . . . not contemptible."[13] Despite such litotes many critics and naturalists recognized the value of Riddell's scientific observations, and twenty-five years later, when the Linnaean naturalist Charles Stewart published a list encompassing the "principal Books which treat of Natural History," *Voyages* was the only text he included by a female author.[14] Riddell's natural history attains this distinction for her empirical reportage on geographical and geological advantages and disadvantages, her detailed accounts of architectural developments, and her minute attention to classifications, descriptions, and uses of indigenous species of the plant and animal kingdoms on the British Leeward Islands. She provides population estimates and precise locations of each island's parishes; scientific descriptions of the soil and minerals in such technical terms as "tetrahoedral chrystals of sulphur," "argillaceous and magnesian earth," and "martial pyrites"; and depictions of the colonies' varying and "beautiful" landscapes (23–24).

In *Voyages,* Riddell often adopts the analytical, objective persona of the naturalist, a mode of writing that conforms to the "new seriousness" of travel literature after 1789, when, as Katherine Turner argues, "the travel writer's nationally representative responsibilities come to the fore, and personal oddities are displaced by the requirements of intellectual and ethical reliability."[15] Women travel writers were in a bind, expected to supply patriotically useful observations and simultaneously maintain an ostensible distance from the masculine sphere of politics. This generic shift is evident in a comparison of *Voyages* with the work of Janet Schaw, a Scottish travel writer who visited the Leeward Islands fourteen years earlier than Riddell.[16] Schaw's earlier account addresses many of the same concerns as Riddell's—especially British nationalism, anti-Catholicism, Protestant industry, and improvements within West Indian nature—but the conventions of her time allow Schaw to be more straightforward in advocating West Indian interests. Riddell more tacitly couches her opinions within "science, minute observations, [and] accurate description," the subtleties of which reveal a complex commentary on nationalism, responding to a particular historical context.[17]

Riddell wrote *Voyages* at a critical juncture in the history of British interests in the West Indies. Arriving on St. Kitts in 1788 and publishing her work in 1792, she recorded her account during the aftermath of Caribbean involvement in the American Revolution and in the midst of tensions and exuberance emanating from the French Revolution prior to the Terror. During the American War of Independence, the sense of national identity predominating in the British West Indies differed from that which spurred revolution in the North American colonies. The majority of West Indians sent their children to Britain for education, many West Indians hoped to return to Britain after making their fortune, and the islands had relatively little communication among them or sentiments of affiliation. British West Indians persistently reiterated their allegiance to Britain and identified with British culture and society rather than establishing a separate Creole national identity within the archipelagos. Feelings of mutual dependence subsisted between the West Indian colonies and the mother country. Indeed, George III considered the West Indian colonies so indispensable during the American Revolution that he "thought it better to risk an invasion of England than to lose the sugar islands, without which it was 'impossible to raise money to continue the war'."[18] However, after the American War, the British government strengthened the Navigation Acts, restricting the islands' trade for vital provisions with the newly independent United States. Abolitionist propaganda campaigns and lobbying from the mercantilist system also threatened the economic viability of the islands, causing unrest among West Indian planters and leading British loyalty to wane. A young Horatio Nelson, sent to the Leeward Islands to enforce the Navigation Acts, stated that the now disgruntled West Indians seemed to him "as great rebels as ever they were in America."[19] In this strained political atmosphere British perceptions of the West Indies were mixed at best. Aware of this, Riddell assumes an appearance of disinterested assessment even as she conveys nationalism through natural history and by manipulating the georgic legacy.

Re-imagining Originality and Hybridity in the "West-India Georgic"

Engaging with subjects typically considered masculine, Riddell integrates into her travel writing the traditional poetic form for establishing science's political and national as well as global influence, the georgic, and its inherent associations with textual hybridity and originality. Scholars often note the disappearance of the georgic after 1767, yet, in many ways, this poetic form is arguably disseminated into the scientific literature practiced in succeeding decades.[20] While the georgic conventionally communicates new scientific thoughts, plans, and discoveries, this pedagogical poetry also sometimes incorporates the formal hybridity of explanatory prose notes, a technique surviving into the natural history poems of late eighteenth-century writers, including Charlotte Smith and Erasmus Darwin. Throughout her travel narrative, Riddell establishes a sense of intertextuality, referencing naturalists and quoting poetic passages from Homer's *Odyssey,* Virgil's *Georgics,* Edmund Waller's "The Battle of the Summer Islands," and James Thomson's *The Seasons.* Although Homer's and Waller's poems are not technically georgics, she chooses lines from these texts that fit easily into the georgic tradition. Interestingly, by dispersing these poets' verses within her informative prose, Riddell arguably inverts the georgic's formal structure to create a related textual hybridity. In other words, these poetic lines act analogously to notes to her prose observations and technical scientific details, providing both explicit and implicit elaboration through their recognizable cultural histories. Each of these poems conveys specific national ideals that become important for Riddell's narrative. Moreover, another georgic forms a powerful backdrop for her text, even in its absence.

Although Riddell does not directly mention James Grainger's georgic *The Sugar-Cane* (1764), which focuses on Antigua and St. Kitts, islands at the center of her work she arguably constructs her narrative with this preceding Caribbean text's techniques for originality, as well as its fate, in mind.[21] Published twenty-eight years before Riddell's *Voyages,* Grainger's poem combines didactic verse and scientific notes to describe the natural history of the West Indies, especially St. Kitts, and the processing of cane into sugar, while also offering advice about the purchase, care, and treatment of African slaves. In his preface he emphasizes the "novelty of the subject," explaining, "the face of this country was wholly different from that of Europe, so whatever hand copied its appearances, however rude, could not fail to enrich poetry with many new and picturesque images."[22] Privileging instruction over pleasure, Grainger points to forerunners in the georgic, such as Hesiod, Virgil, Philips, and Dyer, to excuse his use of "terms of art" connected with agriculture and industry, here those of sugar production. Additionally, he anticipates the objections of Anna Barbauld and other critics who would disapprove of words "not common in Europe," providing notes, for instance, for vegetation such as the ceiba, guava, shaddoc, and sabbaca, and acknowledging, "an obscure poem affords both less pleasure and profit."[23] As he and his readers knew, "profit" signified a chief concern associated with these islands. Grainger himself was a Scottish physician who owned land on St. Kitts and married into "the first family of these islands."[24] In his poem, he laments the death of Riddell's great uncle Ralph Payne, the chief justice of St. Kitts, and calls him his "friend."[25] The poet-physician derived a main source of income from attending to the health of slaves on wealthy estates on St. Kitts and Antigua, and he published *Essay on the more common West-India Diseases* (1764), detailing the illnesses and recommended treatment of slaves, as well as the medical values of the islands' native vegetation; and plants' medicinal virtues also form a particular interest in Riddell's travel literature. A transplanted Scot, Grainger, like Riddell, embodied a national hybridity that is reflected in his poem and its relation to the georgic tradition.

In its origins and themes, the georgic genre represents poetry of nation-building and improvement that became integral to British iden-

tity. As Rachel Crawford explains, the prevalence of the "English georgic" was "closely associated with the Act of Union with Scotland in
1707, which realized the notion of Great Britain."[26] James Thomson, author of the most popular English georgic of the eighteenth century, *The
Seasons,* was in fact Scottish, and emphasized this new Britishness in his
verse. In his "West-India georgic," Grainger carefully refers to "British"
imperial dominance and the steadfast "depend[ence]" and loyalty of
Britain's Leeward colonies.[27] He portrays the colonies as conforming
to georgic ideals, revealing how scientific discovery as well as agricultural and industrial labor in the West Indies contributes to British superiority. However, as several critics have noted, Grainger's celebration
of labor on West Indian plantations maintains an uneasy relationship
with slavery. On the one hand, the poem's speaker expresses a desire to
"knock off the chains/Of heart-debasing slavery," praising the planter-
protagonist for recognizing African slaves' humanity and treating them
with sympathy.[28] On the other hand, Grainger compares slaves to livestock, offers advice about purchasing "negroes" from various regions of
Africa, classifies them according to the kind of work to which they are
best suited, delineates bodily dangers faced by slaves during sugar production, and deems necessary the infliction of occasional violence, stating, "some I've known, so stubborn is their kind,/Whom blows, alas!
could win alone to toil."[29] As Richard Frohock points out, the "georgic is
ultimately a flawed form for celebrating the planter-hero" because while
"it allows for the elevation of agriculture to the heroic work of empire in
the fashion of Virgil's model," it "requires didacticism, and in detailing
the planter's modus operandi the abuses and brutalities inherent in the
chattel slavery system necessarily appear."[30]

Although *The Sugar-Cane* was well received at the time of its publication, British readers, perhaps feeling their hypocrisy in, for instance,
Thomson's famous declaration that "Britons never will be slaves," became increasingly hostile toward Grainger's poem and toward sugar
production in the West Indies as the abolitionist movement gained trac-
tion in the 1780s.[31] By the time Riddell wrote her *Voyages,* the poem's
unpopularity doubtless influenced her own authorial choices. In her
text, Riddell directly references the georgic tradition and incorporates
its conventions, but, probably due to Grainger's troubled depiction of
slavery, she aligns the West Indies instead with "the most fundamental
English georgic of all time," Thomson's *The Seasons.*[32]

Significantly, while Riddell strategically applies the verses of Homer,

Virgil, and Waller to a foreign island (as I will discuss later), she solely reserves those of Thomson, the bard of quintessentially British values, for her depiction of the West Indian colonies, employing his words to begin and end her natural history of Antigua. By quoting Thomson, Riddell frames her portrayal of the Leeward Islands with recognizable, British nationalism. James Sambrook calls Thomson "a child of the Union and perhaps the first important poet to write with a British, as distinct from a Scottish or English, outlook."[33] Moreover, Tara Ghoshal Wallace explains, "While *The Seasons* articulates the poet's discomfort with the practices of the imperial project—its enslavement or oppression of other races, its use of brute military force, its crude accumulation of wealth and political power—the poem tames these brutalities within the domestic space," making the unfamiliar familiar.[34] This tactic of domestication and familiarization is repeated in Riddell's depiction of the West Indies, and she excerpts verses from Thomson's "Summer" as the epigraph of her section describing Antigua:

> To me be Nature's volume broad display'd,
> And to peruse its all-instructing page,
> Or, haply catching inspiration thence,
> Some easy passage, raptur'd, to translate,
> My sole delight. (47)[35]

In Thomson's poem, these lines punctuate Protestant praise of God, leading to this search for divine truth through nature and the poet's art. Riddell thereby claims the poet's role as arbitrator, translating nature and its significance for her readers. Constructing a botanical dictionary of West Indian plants at the end of her travel narrative, she again quotes Thomson, closing her text by stating, "Thus spring the living herbs, profusely wild,/O'er all the deep green earth, beyond the pow'r/Of botanists to number up their tribes," following a description of plants for which, she notes, "The Linnaean Names . . . are unknown," including the *pomme rose tree,* the *conque nut,* and the *bell bush* (105).[36] Riddell thus asserts her work as not only expounding on what is known about the nature of these British colonies, but also as providing information that is new and "unknown," contributing to the British imperial and empirical project. She reinforces this nationalist approach to science through a further intertextual move, applying Thomas Pennant's system of natural history to these West Indian islands.

Pennant's "British" Natural History

A combination of nationalism and feminine propriety arguably underpin Riddell's choice to adopt the zoological system of the British nationalist naturalist Thomas Pennant rather than that of the Swede Carl Linnaeus in classifying Caribbean nature. Although she employs Linnaeus's binomial nomenclature, his "generic and scientific names," she orders plant species alphabetically rather than following his sexual system of botanical taxonomy, and she dispenses with Linnaean zoology altogether, employing instead Pennant's *British Zoology* (1768–70) and *History of Quadrupeds* (1781) to guide her classifications of animals, birds, and fish.[37] In her words, Riddell privileges Pennant's zoology over that of Linnaeus due to the "simplicity" of Pennant's arrangement "as being more elegant and perspicuous" (vii). Pennant himself advertises his system in these terms to contrast with his depiction of the unreliable classes of Linnaeus that change from one edition to another, so that any "Naturalist ran too great a hazard in imitating his present guise."[38] Pennant rejects several of Linnaeus's classes, refusing to endorse him, for instance, in "rank[ing] mankind with *Apes, Monkies, Maucaucos,* and *Bats.*"[39] Through this scientific opposition, Pennant assumes a stance of moral high ground, furthered by his omission of detailed anatomy in describing animals.[40] At a time when some contemporaries balked at women's study of plants' reproductive organs for Linnaean classification in botany, such anxiety was magnified in the realm of zoology, where anatomy more closely resembles that of humans. These considerations of propriety doubtless increased the attraction of Pennant's zoological system for women writers such as Anna Barbauld and Charlotte Smith, as well as for Riddell, who disclaims her text's "accuracy" and "scale" in accordance with conventions of feminine modesty. Yet despite her caution, Riddell's scientific assertions later provoked the satirist Charles Kirkpatrick Sharpe to taxonomize her as "a profligate woman," classing her within her "museum" as a "pickled frog (for such she looked, amid her own collection of natural curiosities)."[41] Generally considered beautiful, Riddell died at the age of thirty-six, so this critical depiction more prominently reveals contemporary hostilities often faced by women daring to address politics and science. Sharpe's zoological comparison associates her educational prowess with froggish (French) Jacobin pursuits of women's rights, displaying her era's political tensions and Britain's fragile national sentiments, of which both Riddell and Pennant were acutely conscious.

Exhibiting nationalism in his popular travel literature, Pennant familiarized readers with the separate regions of Britain, emphasizing the nation's unifying culture-in-formation. He details Scotland's landscape and customs, as well as those of Wales and England, and notes in his autobiography his disappointment in failing to publish his tour of Ireland.[42] Pennant himself was Welsh, and perhaps this shaped his presentation of Britain as a cohesive nation even as he appreciated its individual regions. He was not blind to the divisions that existed but saw his work as an effort to overcome those divisions, urging what Evan Gottlieb terms "sympathetic Britishness."[43] Pennant describes his motivations to write *Tour in Scotland,* stating, "I labored earnestly to conciliate the affections of the two nations [England and Scotland], so wickedly and studiously set at variance by evil-designing people."[44] Matching his words with his deeds, in an appendix to his autobiography, dated 1792 and titled "My Last and Best Work," Pennant recounts that "the dangerous designs of the French" induced him to "form an association for the defence of our religion, constitution, and property, after the example of some of the English counties, cities, and towns."[45] His "zeal" thereby brings Wales into alignment with the British national cause. While recognizing the diversity of British inhabitants and natural resources, he presents a feeling of commonality, especially through differentiation from (and superiority to) continental Europe.

Pennant vies for a quintessentially British system of natural history to which all succeeding taxonomies must trace their lineage, declaring his indebtedness to "our illustrious countryman" John Ray, the late seventeenth-century virtuoso, and crediting him as "the founder of systematic Zoology."[46] Targeting the titans of natural history, Pennant sets up a rivalry between himself, Linnaeus, and Buffon: ultimately a national rivalry between Britain, Sweden, and France, or rather, between Britain and the rest of Europe. The national contest that Pennant stages is as much about economic supremacy as about arrangement of particular orders of animals. He assures readers that British natural resources "give us the superiority over these so much boasted productions of *Sweden,*" and conjures up Britain's "natural" enemy to exhort, "if we reflect but a little on the unwearied diligence of our rivals the French, we should attend to every sister science that may any ways preserve our superiority in manufactures and commerce."[47] Equally paramount to a British system of zoology, Pennant claims a British foundation for natural theology, emphasizing the national merit of Protestant industry.[48] His anti-Catholicism and Francophobia construct a model of national unity forged through opposition to the foreign "other" that resonates with Linda Colley's famous thesis, explaining that Britons "define[d] themselves as Protestants struggling for survival against the world's foremost Catholic power."[49] Indeed, Pennant points to his nationalistic "hints towards enlarging and improving our manufactures and agriculture" as justification for offering his zoologies to the public.[50]

Through the nationalism framing Pennant's system, Riddell imposes these British standards onto the Leeward Islands in order "to conciliate the affections of" not only the different regions of Britain, but also and more pertinently of Britain and its West Indian colonies, without a direct statement of political involvement.[51] Adopting Pennant's national ideology in addition to his classification system, Riddell similarly creates an opposition to the Catholic "other."[52] The first stop in her *Voyages,* the Portuguese island of Madeira, located off the North African coast in the North Atlantic, functions as Riddell's most obvious foil for the British colonies. It is on the Catholic Madeirans that she projects degenerative characteristics often imputed to West Indians.

Madeira's Anti-Georgic

Writing about her visit to Madeira, Riddell correlates the population's prominent Catholicism with corruption and lack of industry. She describes the Madeirans, and especially the lower classes, as not only excessive in their passions, but also as "indolent, dirty, and much addicted to theft" (15). For her, this national character materializes in the disrepair and disorder of the island's main town and in the inhabitants' neglect of nature. Admiring the island's lush vegetation, Riddell states that if Madeira were "properly cultivated" it "might justly be termed the garden of the world." She remarks that "the serenity of the climate, the fertility of the soil, every thing conspire to render it an absolute terrestrial paradise; *and it only requires the nurturing hand of art* to give the finishing touches to a scene on which nature has so profusely poured her choicest treasures" (8, emphasis mine). The failure of the Portuguese to cultivate and instill order in nature provokes Riddell's national indictment of their idleness and degradation.

While Riddell later presents the British West Indies as potentially containing English georgic ideals, Madeira represents the space of an anti-georgic. She strategically employs verses from Edmund Waller and Virgil as epigraphs for her description of this Portuguese island. Riddell first excerpts Waller's lines "The gentle spring which but salutes us here,/Inhabits there, and courts them all the year" from "The Battle of the Summer Islands" (1645), a poem about Bermuda that colors her subsequent Latin quote from Virgil's *Georgics,* translating, "Here the spring is perpetual, and the summer shines in unusual months" (3).[53] Although these quoted passages seemingly praise Madeira as a tropical paradise, her reference to Waller's poem signifies a veiled critique. As A. D. Cousins states, "The Battle of the Summer Islands" is mock-heroic, dramatizing "the incongruity [Waller] posits between the riches of the island and the uncouthness or meanness of its inhabitants" who misuse nature's bounty.[54] Here, "despite an exceptionally benevolent climate and environment, the Golden Age is parodied rather than re-stored."[55] In his classic study, *The Georgic Revolution* (1985), Anthony Low explains that, for many writers of the seventeenth century, the georgic constituted a "shameful" genre due to its "ungentlemanly" associations with labor and agriculture, associations that would undergo positive transformations in the late seventeenth and early eighteenth centuries.[56] Low exemplifies Waller as voicing these cavalier, anti-georgic, antilabor sentiments. Indeed, Waller's dissociation from agriculture forms part of his humorous parody of the Bermudas colony, so that Riddell's second quote from his poem states, "Ripe fruits and blossoms on the same tree live,/At once they promise what at once they give" (9). For Riddell, nature's prolific production on Madeira, as on Waller's Bermuda, highlights the lack of meaningful labor practiced by its "uncouth," "mean" inhabitants.

Recounting that "this desirable island" was first "discovered by an Englishman" and later "conquered by the Portuguese, who set fire to the forests . . . [giving] the soil that degree of fertility which it boasts of at present," Riddell indicates that the Portuguese now have degener-ated from their conquering, industrious past (9). In eighteenth-century imperial rhetoric, this present neglect of the island's natural potential constitutes a forfeiture of land rights that justifies British reclamation.[57] Although not calling for military action, Riddell's observations lend ideological support to Britain's economic domination of the island. Only three hundred people formed the British community of Madeira, as opposed to the more than sixty thousand Portuguese residents. However, this small British population controlled the vast majority of the island's revenue, chiefly owning vineyards and acting as wine merchants to capitalize on the popular wine that took its name from the island.[58] According to Riddell, the British vineyards pleasingly and diligently organized the otherwise chaotic landscape, moving her to quote from Homer's *Odyssey:* "Here order'd vines in equal ranks appear,/With all th'united labours of the year" (10). Regulating both nature and passions, British industry counters the indolence Riddell notes in the island's Portuguese inhabitants and religious representatives, who, in her depiction, by failing to improve nature, become part of it.

Riddell's anti-Catholic critique acquires a satirical dimension through overlapping terminology and methodology between her descriptions of zoological and vegetable "orders" and those of Portuguese monks and nuns. Drawing attention to despondent nuns who envy her freedom at the Convent of Santa-Clara as well as to a monastery of Franciscans, she humorously details the characteristics of monks in this "strict order" as minutely as those of biological species in her natural history (5, 7).[59] Given her witty affect, she may have read the anti-monastic satire *John Physiophilus's Specimen of the Natural History of the Various Orders of Monks, After the Manner of the Linnaean System,* pseudonymously authored by Edler von Ignaz Born, and printed in London for Joseph Johnson in 1783. In Born's ludicrous treatment of what he deems "the study of Monkhood" or "Monachology," he declares his discovery of "a genus entirely new . . . I mean the monk: a genus most unlike the human, yet belying the human form."[60] This ribald text uses "Monks" synonymously with "insects" when quoting the authority of Linnaeus. According to Born, the "genus" of "The Monk" is also "an animal greedy, stinking, filthy, thirsty, slothful, preferring hunger to labour. At the rising and setting of the sun, but especially at night, the monks flock together, and when one begins, they all set up a howling: They all run together at the sound of a bell."[61] Although Riddell's more serious work precludes extensive ridicule, her description of the Franciscan order closely compares with that of Born, who remarks on their "indolence" and hypocrisy. After delineating the Franciscans' physical characteristics—"the Friars go barefooted; their habit is a brown, coarse stuff, with a cowl; and they have a cord tied round their waist," and so on—Riddell relates that "all sorts of property are forbidden by this rule . . . they are also forbidden to receive money" (7). Like Born, she records the monks' hollow virtue, stating, "I had a very entertaining proof of their ingenious equivocations in that respect; for, on offering one of them a piece of money, he held up his hands as if fearing they should be polluted by the touch, at the same time turning his head, pointed to a little pocket, in which I accordingly slipped the dollar" (7). While wryly deriding the monk's "ingenious equivocations," Riddell presents such minor failings in Catholicism as symptomatic of more sinister corruptions.

Riddell's observations on a Catholic funeral on Madeira anticipate Vincent Brown's recent claims that rituals for the dead in the Atlantic world reveal broader cultural values.[62] At the funeral, Riddell relates that the bones of the corpse, "as soon as it was brought within the church . . . were all broken one after another, the body carelessly thrown into the ground without a coffin, and the hole filled up with large stones" (15). This grotesque Catholic burial "did not a little disgust and surprise" Riddell and would have reminded her readers that Protestants were permitted no church on Madeira, nor burial until 1767. Before that time, Protestant corpses were unceremoniously dug up and thrown into the sea.[63] In light of such depravations, Riddell later wrote, "I . . . think unfavourably of man only as ill-organized civil societies . . . and false religions, have degenerated him."[64] For her, both their "indolen[ce]" and unimproved land reflect the religious and cultural corruptions of the Portuguese on Madeira; combined, they create a standard of foreign degeneration against which the British West Indies may be measured.

NATIONAL/NATURAL HYBRIDITY

In addition to its textual, formal hybridity, *Voyages* conveys British national hybridity through description of the West Indian colonies. Upon reaching the Leeward Islands and disembarking from her aptly named ship, the *Britannia,* Riddell records her observations on St. Kitts, Nevis, Barbuda, and Antigua. In their "vast rocks, high precipices," "deep vales, and hanging woods," "rich pastures, grazing cattle, and little gardens dispersed on the slopes of the hill," the islands' natural scenes have all the diversity—and familiarity—of Britain itself (24). Riddell attends to the islands' incorporation of different British regions through both the Britannic naming of places and the representations of regional populations within the island communities. For example, using local dialect, she refers to two disparate terrains of Barbuda as "the High-lands" and "the lowlands," so that, at times, one almost forgets that she is not speaking of Scotland (36–37). Again situating Scotland within the West Indies, Riddell describes the expedition of "three hardy Scotsmen" who planted a flag near a mountain's summit on St. Kitts in 1787, a conventional symbol of claiming land in the name of one's nation (22). In these observations she subtly echoes the tendency of British writers to find (or create) elements of Britain in its colonies. Janet Schaw, for instance, wrote of St. Kitts "that its principal beauty to me is the resemblance it has to Scotland."[65] Riddell's narrative highlights further regional denominations, such as "Irishtown" on St. Kitts and "English Harbour" on Antigua, which "is not open to any craft but what belongs to the King," giving the islands a strangely British feel despite their exoticism (51). She also conjures up images of national unity forged through opposition to the foreign "other." For example, Riddell invokes "the celebrated fortress of Brimstone-Hill" on St. Kitts, where colonists put up the bravest resistance of the American War in the Caribbean theater by holding out against a French naval siege for five long weeks in 1782 (28). She thus revives memories of colonists recent loyalty and unification against French forces in the British national cause. As in Pennant's works, Riddell's inclusion of Britain's diverse domains retains an appreciation of difference while producing a capacity for hybrid national identity. Her *Voyages* evinces precedence for this *national* hybridity within the *natural* hybridity of the Leeward Island ecologies.

Riddell exhibits a naturalist's fascination with organisms that appear to bridge gaps between species, orders, or even kingdoms, such as the "*sea bat* or *laphius vuspertilio,* which, though an inhabitant of the ocean, carries in its form the striking resemblance of an ill-formed quadruped," as well as "three species of the *ascidia,* or *animal flower,*" and the opossum, which in her description simultaneously incorporates the anatomy of a weasel, hog, and "domestic cat" (55, 68, 78). Often denoted in species' common names, this kind of hybridity also appealed to eighteenth- and nineteenth-century poets who imaginatively explored natural objects' visual analogies and surprising liminalities.[66] Through astute study of the islands' natural history, Riddell understood that present biological species did not necessarily indicate long-term habitation any more than did current colonial occupants of these islands that successively (and sometimes simultaneously) belonged to Spain, France, and/or Britain, exhibiting a history of national hybridity.[67] As the Leeward Islands were never attached to a mainland, species existing on these islands resulted from dispersal over water, a fact acknowledged by Riddell through her delineations of feral species, explaining that the Spanish in the fifteenth century stocked the islands with cattle, goats, and hogs.[68] Thus, in addition to awareness of origins, she demonstrates knowledge of how political and natural histories of the islands intertwine.

Moreover, Riddell records species differentiations or changes in biological forms among the islands similar to those that prompted Darwin and Wallace to recognize that islands act as "evolutionary laboratories." Although Riddell primarily confines herself to the natural history of Antigua, she makes a point of noting additional species she believes to be unique to St. Kitts, especially focusing on one species of lizard in which the head "is always of a bright flame colour, and very beautiful: I never knew of this particularity being observed among lizards in any of the Caribbee Islands except St. Kitt's" (29). Of interest to recent biological researchers, the lizard populations of the Greater and Lesser Antilles have become the modern equivalent of "Darwin's finches" as a means to understanding speciation, and "since archipelagos are forums for recent adaptive radiations, and hybridization occurs in the early stages of the differentiation of a taxon, an unusually high incidence of hybridization on islands may in fact occur."[69] But Riddell's understanding of biological hybridity and originality, of course, engages with more contemporary discussions of reproduction and climatology. She emphasizes, for instance, the importance to the islands of their thriving population of mules, which are "extremely serviceable," "much employed," and "bred and imported here in great numbers" (53). In the late eighteenth century, conceptions of the Caribbean as especially conducive to hybridity had their source in theoretical debates of natural history in which "mules"—a term that both designated the progeny of a horse and an ass, and could be used interchangeably with "hybrids" and "mongrels" at this time—played a central role.[70]

Hybridity informed the naturalist Buffon's understanding of zoological species, as well as of the West Indies, as demonstrated in Smellie's translation of his work. For Buffon, degeneration occurred largely through the effects of food and climate over time and, by means of it, "a distinct species" could emerge.[71] Distinction among species could be tested by causing a "degenerated" individual to mate with a member of its original "family" and, if the intermixture proved either barren or produced hybrids, the species were indeed distinct. Buffon vehemently asserts that zoological hybrids occur most frequently in warm climates and refutes the widely accepted conviction of mules' absolute sterility.

Significantly for the work of Riddell, he documents the reproductive capacity of a "she-mule" on St. Domingo, a colony of the French West Indies, to corroborate his insistence that warm regions promote hybridity.[72] He also indicates the possibility of considering races analogously with hybrids: "Does not a race, like the mixed species, proceed from an anomalous individual, which forms the original stock?"[73] Here, Buffon poses a question that interchanges concepts of hybridity and originality. While maintaining that humans compose a single species, he implies that climate's ability to originate distinct "races and nations" is increased in warm regions, resulting from the increased production of hybridity. Contemporary theories of natural history therefore endowed the West Indies with a hybrid potential that Riddell could invoke while portraying a hybrid British nationality.

Buffon famously invested the Americas with heightened potential for alteration, mainly conceiving this possibility in degenerative terms. Degeneration was often manifested in a diminishment of reproductive power, as well as of size and strength in animals and of activity in humans. He saw degeneration as particularly exemplified in the American Indian. Such assertions of degeneration provoked Bryan Edwards, in his history of the British West Indian colonies, to denounce "the speculations of Mons. Buffon and some other French theorists" who posit that "the New Hemisphere" decreases "the capacity of improvement" in the human species.[74] Yet Buffon also provides means for positive renovation in the Americas. He admits that several species of quadrupeds "have improved by the influences of the [American] soil and climate."[75] Buffon notes that "degeneration" and "improvement" are essentially the same in the view of "Nature" because both indicate a change from the original form.[76] He sometimes oscillates in value judgments of alterations, especially in instances of domesticated animals whose "improvements," as perceived by humans, he deems more likely to be degenerations for the animals as species.

Riddell similarly interrogates subjective valuations of "improvement" and "degeneration." She records changes occurring in zoological species transported from Europe to the West Indies, observing, for example, that "the sheep soon lose their woolly fleece in this climate: Providence has clothed them with a lank brown hair instead; which, though it diminishes their beauty, is infinitely more serviceable to them in point of coolness" (54). At a time when species fixity was the orthodox doctrine of the Church and of a majority of naturalists, Riddell mitigates her straightforward assertion of species mutability by imputing these changes to God's wisdom. She also enters the Buffonian conundrum of questioning whose perspective matters when labeling a species "degenerate." The sheep may have degenerated according to European standards of beauty, but Riddell presents the species as improved in terms of making a successful adaptation to the environment that will increase its ability to survive and propagate.[77] As my next two chapters show, such debates about relative improvement and degeneration, especially in determining what constitutes originality and hybridity, became important in this era with regard to literary as well as biological forms. Interestingly, in their respective texts on the West Indies, Edward Long and Bryan Edwards correspondingly assess climatic alterations in the anatomy of British colonists, who, they claim, have developed deepened eye sockets that conveniently provide shade from the "strong glare" of the sun.[78] For Riddell, improvements in the British colonists of the Leeward Islands proceed from their interactions with nature. Just as the environment is conducive to promoting British national hybridity, she indicates its ability to promote British values.

BRITISH VALUES IN THE WEST INDIES

British Protestant industry supports a relationship with nature that eighteenth-century philosophers molded to fit the northern climate, portraying adversities in nature as a blessing that produces a nation of improvers. According to Buffon, "there is a direct correlation between the degree of civilization that a given people attains and the mastery that they exercise over nature."[79] In his *History of Jamaica* (1774), Long emphasizes that sterile land increases the industry and ingenuity of its inhabitants.[80] This idea is further supported in Crèvecoeur's *Letters from an American Farmer* (1782), which theorizes, "Where barrenness of soil or severity of climate prevail, there she [Nature] has implanted in the heart of man sentiments which overbalance every misery and supply the place of every want," while "extreme fertility of the ground always indicates the extreme misery of the inhabitants!"[81] In the midst of these contemporary assertions, Riddell plays down the paradisiacal conception of West Indian nature, instead focusing on the islands' natural adversities as well as the colonists' ability to overcome these adversities and even convert them to advantages.

Enlisting corroboration from Thomson's *The Seasons,* Riddell's depiction of the West Indies suggests British georgic ideals of Protestant labor and nation-building improvements. In contrast to the indolence indicted in Madeira, she records numerous examples of industry within the Leeward Islands. Riddell expresses admiration for the islands' architecture, remarking on its utility and stylish modernity. Unlike the main city in Madeira, that in Antigua "is one of the largest and most handsome in the West Indies. The streets are wide and well laid out and the houses mostly commodious and airy"; the city's church, courthouse, and military barracks are "elegan[t]" and "handsome" (49). Moreover, while the soil of these islands is not presented as boasting the richest prospects, the taxing nature of the British West Indies serves as a guarantee that its inhabitants will prove industrious. Riddell describes the salt lakes of St. Kitts, for instance, as foul-smelling and noxious, but they are also deemed very lucrative (27). On Antigua, the scarcity of fresh water necessitates construction of great "tanks and cisterns" to preserve rainwater for its inhabitants, and "leaden pipes" have been structured for transference of spring water on St. Kitts (48, 26).

In being able to control and improve the treacherous environment of these islands, Britons (im)prove the worth of their nation by increasing the trade and prosperity of the British empire. Riddell highlights several indigenous species of poisonous plants, lizards, and insects, but can sometimes counter with a vegetable antidote found on the islands. Her botanical observations delineate a litany of manufacturing, and especially medicinal, uses that suggest the West Indian propensity for disease as well as the ability of British scientific discoveries to nullify this threat. Recasting the colonies as producers of medicine (rather than of disease), Riddell presents "castor oil," for example, an "almost infallible medicine in cases of the greatest danger," as "one of the most valuable tropical productions imported to Europe" (102). She often stresses the natural dangers of the islands' coral reefs and the difficulty of travel on certain terrains that require excellent piloting skills. These natural dangers serve both as a natural defense for the islands and also as a natural test of superiority. In a similar vein, Edward Long noted the greater natural fertility of soil in the French West Indies yet credited West Indian Britons with the capacity to overcome deficiencies in their own islands and to dominate sugar production.[82] The manufacturing of sugar and the slavery that enabled it, however, mark a facet of British industry that Riddell largely ignores.

SLAVERY AND THE PARADOX OF HYBRID BRITONS

Considering the overwhelming predominance of slavery in "the sugar islands," Riddell's refusal to treat this fundamental aspect of West Indian society produces a striking absence in her narrative. Certainly for those sharing the anxieties of Edward Long, "the father of English racism," the West Indian potential for hybridity immediately conjured up slave-owners' unruly passions, resulting in taxonomies of racial gradations in the islands' inhabitants.[83] Long conflated the word "mulatto" with naturalists' derogatory perceptions of "mules" or hybrids, and warned that such racially "mixed progeny" would form "a vicious, brutal, and degenerate breed of mongrels."[84] At the same time, abolitionist debates raging in Britain in the 1780s and early 1790s associated brutally corrupt degenerations in the West Indies not with Long's so-called "mongrels," but with slave-owners whose actions caused many Britons to feel that the integrity of their national character was at stake. Abolitionists insistently posed the question, How can a nation that prides itself on, and indeed, *defines* itself by, its freedom permit the atrocity of slavery to occur under its government? In her narrative, Riddell only once denominates "slaves" as such, generally referring to blacks as "Negroes," a term that elides their subjugation (37). She chiefly speaks of their particular uses of plants and animals, which makes it seem as though West Indian slaves are hardly slaves at all, and possess abundant time to devote to their own needs.

Even when the presence of sugar production and slavery is most apparent in her natural history, Riddell shifts focus away from the West Indian colonies' dependence on these controversial institutions. For example, in her alphabetical descriptions of plants found on Antigua, she arranges these vegetable species according to their Linnaean or scientific names, so that "*saccharum arundo,* or *sugar cane,*" is briefly listed alongside various other botanical species, without distinction or recognition of its particular economic viability. Indeed, while offering more detail about the uses of other plants, she simply and parenthetically glosses sugar manufacture as "a long and tedious process," avoiding description of the labor she witnessed forced on African slaves. Additionally, in the single instance where she brings slavery to the fore, she displaces this labor. Portraying the island of Barbuda, she relates, "The inhabitants are black, mostly slaves to Sir William Codrington; their employment is husbandry. The sugar cane does not flourish here in any perfection.

The chief commodity, from which the owner derives any benefit, is the cattle, which breed wild among the woods, and sell to prodigious advantage in the neighbouring Caribbean isles" (37). Removing the possibility of growing sugar cane "in any perfection," Riddell here severs the connection between sugar and slavery typically associated with the West Indies, instead portraying African slaves in the more familiarly pastoral, British "employment" of raising cattle. In her depiction, slaves barely labor in this profitable "husbandry," for the cattle "breed wild among the woods," necessitating little additional care. In this way, Riddell negates the work and brutality of the slave system. Her suppression of slavery instead directs attention toward West Indian colonies' fortifications and urban improvements that champion British industry (with little indication of who actually performs this labor), and the discoveries of various uses for botanical species. While Grainger's *The Sugar-Cane* required georgic attention to the labor, and thus slavery, involved in West Indian sugar production, Riddell's selective reportage, mixing travel literature, poetry, and natural history, allows her to avoid the painful realities of slavery in her narrative.

Riddell's motives for overlooking slavery in *Voyages* were complex. Her family of plantation owners clearly had imbibed the West Indian doctrine of white supremacy, but Riddell herself held strongly abolitionist principles, as is evident in a letter to Smellie that reflects on "the accursed traffic of the slave trade" and the massive and bloody slave insurrection on St. Domingo in the French West Indies in 1791.[85] Although Riddell "deplore[s]" its "effusion of . . . blood," her discussion of the Haitian Revolution reveals her support for "liberty" and "equality" as well as a faith that the West Indies can and will improve toward these ideals, even if by violent means. In *Voyages*, on the other hand, probably partially in deference to her family's sentiments, Riddell sets the danger of insurrection in the distant past, and thereby minimizes this danger's actual escalation as whites became increasingly outnumbered by the slave population (50).[86]

This threat of slave revolt had a paradoxical effect on white colonists' notions of British freedom and identity, one that forms another motive for Riddell's elision of slavery. When Bryan Edwards describes the chief characteristics of these island societies, he particularly lauds their fervent dedication to liberty and equality. He explains that the imposition of slavery on blacks causes white colonists to demand greater equality from other whites, regardless of class or nation, to an extent that would never be allowed in Britain.[87] Here, Scots, Irish, Welsh, and English of all stratifications are on virtually equal footing and, indeed, band together to form a solid front of white supremacy. Thus, for Riddell, addressing slavery in her narrative would have meant having to admit that the national hybridity, the "British" unification found in the West Indies, was due less to an intensification of British georgic values than to perversion of those values and maintenance of the racial hierarchy born out of slavery. Her refusal in *Voyages* to consider this West Indian paradox underscores her embarrassment at the continuance of racial oppression and her recognition of its growing unpopularity with British readers. However, at the same time, her depictions of British industry and of blacks in a state resembling emancipation arguably register Riddell's hopeful vision of improvement in these hybridity-prone islands whose natural history remains charged with so much potential for change.

Emerging ideas about transformations of races, nations, and species and their relation to fixed patterns in nature created great controversy at the end of the eighteenth century. Although Riddell's fluid notion of hybridity and originality assumes these concepts' interchangeability in that hybrid forms can produce something new and original in literature, nature, or nation, this supports one side of a larger contemporary debate that met with fierce opposition. Many literary and scientific writers, including Anna Seward, instead dichotomized notions of origins or originality and hybridity, viewing the latter with suspicion. Returning to England, the next chapter examines such discussions' implications for literary plagiarism and the relative fixity of poetic, as well as biological, structures.

Poetic and Biological Forms

PLAGIARISM, ORIGINALITY, AND HYBRIDITY

3

The Evolution of the Plagiarist

Natural History in Anna Seward's Order of Poetics

Anna Seward's vitriolic attacks on the poetical borrowings of Charlotte Smith remain a formidable embarrassment for scholars seeking to restore the prominence of Britain's two most popular female poets of early Romanticism. While Smith has received notable critical attention in recent decades, Seward is currently best known not for her literary efforts but for denouncing Smith's *Elegiac Sonnets* (1784) as "hackneyed" "hedge-flowers"—"full of notorious plagiarisms."[1] Seward's accusations likely influenced Smith to insert quotation marks in the third and subsequent editions of her sonnets, and the critical power Seward exerted in her lifetime continues to hold sway over modern scholars who repeatedly have been put on the defensive, compelled to find various interpretations of Smith's plagiarisms that might defuse imputations of scandal.[2] Perplexed by the viciousness of Seward's critiques, critics often trivialize them as merely denoting jealousy of Smith's literary success.[3] As most treatments of this controversy center on Smith's vindication, few attempts have been made to assess the motives behind Seward's attacks. I would like to open a new mode of explanation for Seward's anxieties about intertextuality in the form of plagiarism, one based in her conception of natural history.

Significantly, Smith was not the only poet to draw Seward's disapproval for acts of plagiarism. The poet, physician, and naturalist Erasmus Darwin was Seward's neighbor of nearly twenty-five years, and Seward eventually became his biographer, publishing *Memoirs on the Life of Dr. Darwin* in 1804. In the *Memoirs,* Seward justifiably accuses Darwin of plagiarizing lines of her verse in the first part of his long scientific poem *The Botanic Garden* (1791). As I will demonstrate, natural history taxonomies functioned as a template for what I term Seward's "order of poetics" and her perception of plagiarism can best be explained within this larger critical context. Seward grounded much of her authority in the overlapping vocabularies, methodologies, and teleologies of literary criticism and natural history, assuming the critical persona of a literary naturalist.[4] Examining this interrelation will shed new light on current scholarly studies of plagiarism and poetic form, as well as on Seward's censures of Darwin's and Smith's poetry.

Seward's thinking about literary imitation was shaped by a belief in fixed biological forms that tended to force newness into two categories: originality and hybridity. She viewed deviations from principles of originality with distrust and considered Darwin's and Smith's plagiarisms to be degenerate, stylistic hybrids. By analogizing authorial style with fixed species in her critique of Darwin's early theory of evolution, Seward associated unsuccessful literary imitation with zoological mutability, and Darwin's violation of both natural and poetic order applies to Smith's imitations as well. But although Seward condemned Darwin's stylistic hybridity, she praised his formal choices as "original"—an endorsement categorically denied to Smith's sonnets. Seward's derogations of Smith's formal and stylistic hybridity ultimately mark her efforts to disqualify this rival from what Seward called "the sonnet claim."[5] Appropriating the naturalizing authority of science, Seward asserted the originality of her own works while indicting the hybrid monstrosity that made Smith's verse unclassifiable within the order of poetics.

LITERARY NATURALISM

Her modest disclaimers make it easy to overlook Seward's critical engagements with natural history. Referring to the *Memoirs* as "my little Darwiniana" and "my feminine Darwiniana," she employed diminutive rhetorical gestures in anticipation of derisive responses to her criticism of Darwin's scientific texts (6:55, 94). These *Memoirs on the Life of Dr. Darwin* are, in fact, less a biography of Darwin's life than a criticism of his works that brings the focus back to Seward. The text's malicious undertone led Charlotte Smith to be "reminded of a jackal at prey."[6] Nor were reviewers to be put off by Seward's disavowals of scientific pretension. A commentator from the *Critical Review* uprooted Seward from scientific discourse to plant her securely in the realm of literary criticism, remarking that "in her critical examination and analysis of the Botanic Garden, miss Seward is more at home than in ascertaining the comparative merits of the Zoonomia with the works of Hippocrates and Galen."[7] Relegating her to the literary realm reflects a prejudice against her gender with which Seward was well acquainted.

Scoffing at contemporary objections to women's study of natural history due to its emphasis on anatomy and reproduction, Seward defended the propriety of reading Darwin's poems *The Botanic Garden* and *The Temple of Nature*.[8] She rationalized the depiction of "floral harems" in the former as being consistent with the "real," "discovered" Linnaean system of plant sexuality, and asserted that the latter poem "can only be unfit for the perusal of such females as . . . are totally ignorant that, in the present state of the world, two sexes are necessary to the production of animals" (6:84).[9] These justifications accord with Seward's support of female education and her admiration of Wollstonecraft's "wonderful book, The Rights of Woman" (3:117). Still, she understood the need for caution.

When Darwin first conceived of *The Botanic Garden,* he suggested that they divide the task, but Seward declined, believing that "the plan was not strictly proper for a female pen."[10] She later expounded, "That which it might not be strictly proper for a woman to write, may yet be not unfit for her perusal" (6:144). Her cautious double negative maintains women's right to attain knowledge, particularly of natural history, and simultaneously acknowledges ideological restraints necessitating women's prudence when committing such subjects to print. Seward exercises her propriety specifically in regard to Darwin's plan—one can imagine the critical sarcasm and moral censure to which Seward would have been exposed as an unmarried woman collaborating with Darwin in versifying the "sexual system"—but she had no qualms about applying natural history to poetry, per se. She praises this "original" aspect of Darwin's poem, and her critiques of Smith do not mention her use of natural history. Some of Seward's own poems express similar scientific knowledge, and she often insisted on natural-historical accuracy in the poetry of others.[11] For example, in criticism of *The Seasons,* Seward writes that "one of the most strikingly exceptional violations of NATURAL HISTORY is committed by the generally so very accurate Thomson. . . . it is a gross anachronism to attire the SPRING in [roses]" (1:19–20). Regarding a poem by Helen Maria Williams, Seward notes, "Helen is also a little out in her zoology" (3:6); and even Erasmus Darwin receives correc-

tion when Seward, critiquing *The Botanic Garden* in the *Memoirs,* catches
him in a false ornithological observation regarding the redbreast, and
disapproves of botanical alliances formed around the English nettle.[12]
Seward felt a deep conviction in her critical capacities and recognized
that employing science as an underlying framework could add authority
to literary criticism.[13]

In her critical analyses, Seward functioned as a literary naturalist,
closely observing minute details of literary works as the naturalist scru-
tinizes "works" of nature. The rhetoric of late eighteenth-century literary criticism overlapped with that of natural history so that Seward, like
other critics of her day, referred to "species," "orders," and "classes" of
verse. She also obsessively practiced the naturalist's methods of classification: ordering and ranking not only individual poems, but the poets
themselves. In so doing, Seward was in step with her time. Naturalists
of the eighteenth century often practiced a nascent form of what we now
call "sociobiology."[14] And this interaction of science and sociology can
be seen, for example, in William Withering's interpretation of the Linnaean system, where "species resemble individuals," "classes resemble
nations," and so on.[15] Seward reveals her own tendency to naturalize
sociopolitical convictions through correlations with natural history in
a letter published in the *Gentleman's Magazine* in 1793. Disillusioned by
the violence of the French Revolution, she attempts to persuade her
friend and correspondent Helen Maria Williams to return from Paris,
arguing that "the different talents and dispositions of men, inherent and
acquired; the comfort, protection, and prosperity of civilized society; *the
dispensations of providence in the vegetable, animal, and rational universe;* the silent
lessons of natural religion, and the precepts of revelation, are all the reverse of Paine's equalizing creed" (3:203, emphasis mine). In the late
eighteenth century, terms associated with "order" invited social at least
as frequently as natural-historical application. In her letter to Williams,
Seward holds up the ranks and distinctions in nature to corroborate
social hierarchy, a paradigm that, through her literary criticism, she extends to "the different talents and dispositions" of poets.

By arranging poets into a taxonomy resembling systems of natural
history, Seward charts one model of the hierarchical canon that appeared for the first time in English criticism in the second half of the
eighteenth century, and thus concurrently with Britain's rising interest
in natural history.[16] Seward and her contemporaries employed a "logic
of differentiation" that systematized the canon within "workable nor-
mative boundaries" that "could be ever more finely specified."[17] These

canon-makers' attempts to combine absolute valuation with an "open-ended process of comparison" parallels natural-historical efforts in the late eighteenth and early nineteenth century to synthesize the method-ologies of Carl Linnaeus and Georges-Louis Leclerc, comte de Buffon, arguably the two most significant naturalists of the eighteenth century. While Linnaeus emphasized the rigid classification of species based on a minimum number of morphological differences, Buffon viewed Lin-naean classification as too "abstract," insisting instead on "a complex interweaving of behavioral, biological, geographical, and relational properties that set one species apart from another."[18] Seward's order of poetics attempts to synthesize similar, competing methodologies, pro-ducing tensions in her ordering principles.

In Seward's formulation, a rigid, hierarchical chain of being con-verges with an attempt at objective appreciation of each poet's intel-lect or verse specialty, just as naturalists often maintained hierarchical notions even as they strove to analyze each species with attention to unique characteristics, formed to particular behaviors and environ-ments. Seward's letters are filled with these groupings and separations, associating poets according to the influence of literary predecessors, or by their predilection for a given verse form so that each poet belongs to a particular class:

> The first class seems formed by those who are at the head of some particular branch in their science;—as Spencer of the allegoric; Shakespeare of the dramatic; Milton of the epic; Butler of the burlesque; Dryden, Pope, and Sam. Johnson, of the ethic, heroic, and satiric; Thomson of the descriptive; Prior of the narrative and epigrammatic; Gray of the lyric and elegiac; Shenstone of the pas-toral.
>
> Admitting the justice of my criterion for the formation of the first poetic classes amongst our authors, it must yet be confessed, that there are, in the second, bards of more exalted genius than some whose names have a right to be arranged in the first, as being first in their line of writing. For instance, Collins and Mason are much greater poets than Butler and Shenstone; but then they have, in Gray, a superior in their line, the lyric.[19]

By familiarizing herself with the traits of individual poets, Seward felt that she could, in addition to ordering a literary taxonomy, spot imita-tions in the works of others and even identify the authors of anonymous poems.

CLASSIFYING THE PLAGIARIST

When a Linnaean botanist lights on an unknown specimen of plant, he classifies it by identifying singular characteristics that prove its conformity with a class already in existence. Similarly, Seward, thriving in the role of *literary* naturalist, took pride in her ability to identify unknown specimens of verse. But rather than counting stamens, Seward looked to the poetic attributes that would indicate the work of a specific author. In the *Memoirs of the Life of Dr. Darwin,* Seward relates an instance in which Darwin published a poem anonymously, and she boasts the success of her method of classification, having "s[een] the Darwinian stamp on the lines at one glance . . . as if the peculiar style and manner of his muse were not instantly apparent."[20] Seward's identification of "the Darwinian stamp" reflects the notion that each poet possesses a distinct style that is his own individual property and which can be readily identified and classified under that author. She details Darwin's authorial traits as a naturalist might enumerate the characteristics of a particular species: "The Darwinian peculiarity is in part formed by the very frequent use of the imperative mood, generally beginning the couplet either with that, or with the verb active, or the noun personal. Hence, the accent lies oftener on the first syllable of each couplet in his verse than in that of any other rhymist; and it is, in consequence, peculiarly spirited and energetic. Dr. Darwin's style is also distinguished by the liberal use of the spondee."[21] Seward here indicates the ease with which one who is familiar with a poet's style might correctly classify a specimen of that poet's verse. She also later cautions that "we ought to look jealously at all which do not carry to the mind of the reader internal evidence of their imputed origin."[22] This plea for circumspection in verifying authorial identity through internal evidence signifies her anxiety for the correct attribution of her own poetry at least as strongly as that of Darwin. For Seward, being able to recognize a poet's style had important bearings on determining plagiarism.

In the *Memoirs,* Seward recounts the genesis of *The Botanic Garden,* explaining that Darwin's poem grew from verses written by *herself* upon first seeing Darwin's real-life botanic garden in Lichfield in 1779. Darwin declared that Seward's verses should "form the exordium of a great work," and he made the proposal of poetic collaboration that Seward refused. She relates that Darwin then sent her verses to the *Gentleman's Magazine* "in her name . . . but, without consulting her, he had substituted for the last six lines, eight of his own. He afterwards, and again without the knowledge of their author, made them the exordium to the first part of his poem. . . . no acknowledgment was made that those verses were the work of another pen. Such acknowledgement ought to have been made, especially since they passed the press in the name of their real author. They are somewhat altered in the exordium to Dr. Darwin's Poem, and eighteen lines of his are interwoven with them."[23] Underlying Seward's exposé is the conviction that her style, the peculiar quality of her verse, is at stake.[24] As Tilar Mazzeo's recent study makes clear, when original poetry is sufficiently distinct, it "remains so tied to the person of the author . . . that it remains [her] own even in the context of other texts," and imitation is then bound to fail.[25] Thus, for Seward, the distinctness of her poetic voice ensures its recognition and the failure of Darwin's imitation. Because, in this era, imitation did not necessarily preclude originality, determining an attempt's "success" or "failure" was of the utmost importance.

Literary Romanticism concerned itself with two major categories of plagiarism: "poetical" (or "aesthetic") and "culpable." Culpable plagiarism entailed a moral judgment, but charges were rare and extremely difficult to prove since one had to demonstrate that borrowings "were *simultaneously* unacknowledged, unimproved, unfamiliar, and conscious."[26] Poetical plagiarism, on the other hand, indicated an aesthetic judgment that could rest on one or more of these four kinds of borrowings, and such charges were fairly frequent in the Romantic era. Seward's accusations that Darwin plagiarized her verse were often strongly worded; she wrote of the lines of Darwin's exordium, "four-fifths of them are mine verbatim, and mine the whole order of the scenery, so that a charge of plagiarism must rest somewhere" (3:156). However, as Seward knew, Darwin's interweaving of additional lines with her own constituted an attempt at improvement, so that regardless of his degree of aesthetic success or failure, he could only be guilty of poetical (not culpable) plagiarism. Indeed, her accusations of poetical plagiarism against both Darwin and Smith centered on this aesthetic question of improvement. While successful improvement was heavily determined by unity of style, requiring the seamless integration of borrowed texts into the voice or style of one's own production, "unimproved texts were frequently described as monstrous."[27]

Interestingly, Darwin delineates his own conception of plagiarism in the third interlude of his *The Loves of the Plants* (the second part of *The*

Botanic Garden, which was published before the first, in 1789). In a move that struck Seward as rife with hypocrisy, Darwin explains that "perhaps a few common flowers of speech may be gathered as we pass over our neighbour's inclosure, without stigmatizing us with the title of thieves; but we must not therefore plunder his cultivated fruit."[28] Admitting his conscious imitations of Edward Young's *Night Thoughts* (1742) and John Langhorne's *Country Justice* (1774), Darwin goes on to state that "there are probably many others, which, if I could recollect them, should here be acknowledged. As it is, like exotic plants, their mixture with the native ones, I hope, adds beauty to my Botanic Garden."[29] By discussing plagiarism using the language of vegetation, Darwin participates in the conversation of organic originality begun in Young's *Conjectures on Original Composition* (1759). If not for its preceding date, Young's work could almost read as a deprecating response to Darwin's claim for the "beauty" of mixing other poets' "exotic plants" with his "native ones," for Young contends that "an Imitator is a transplanter of Laurels, which sometimes die on removal, always languish in a foreign soil."[30] Despite Darwin's plagiarism of her verses, Seward often praised his originality, but it was an originality that she found compromised. Throughout the *Memoirs,* Seward displays examples of Darwin's imitations, both literary and scientific.[31] Evaluating instances of plagiarism was crucial to literary criticism in the decades surrounding the turn of the century, and Seward's letters abound with casually mentioned detections of borrowings in the works of Virgil, Spenser, Shakespeare, Milton, Dryden, Pope, Swift, Sterne, Gray, Chatterton, Cowper, Burns, Southey, and others. For Seward, literary imitations were a matter of course that could greatly enrich poems when successful, or create monstrosity when trespassing against the critical guidelines of plagiarism. Therefore, each instance had to be considered individually to determine its legitimacy. In the *Memoirs,* Seward adopts Darwin's technique of discussing literary imitation through the rhetoric of natural history, and monitors plagiarism's chaotic potential when analyzing his most famous scientific work, *Zoonomia* (1794–96).

While Darwin imbued *The Botanic Garden* with hints of what we now call biological evolution, it was in *Zoonomia* that he declared his faith in "perpetual transformations" of species. Darwin began writing *Zoonomia* in 1770, and his preoccupation with evolution can also be traced to this year. Seward records that in 1770 Darwin painted on his chaise "his family-arms, which are three scallop-shells," accompanied by the motto he inscribed to it, "*Omnia e conchis,*"—"Everything from shells" (6:136–37). Seward's father, the Canon of Lichfield Cathedral, wrote a "satirically-playful epigram" on the subject that induced Darwin to paint over the arms and motto. Thomas Seward voiced one of the two main objections that would bring Darwin, *Zoonomia,* and these early evolutionary ideas under more widespread attack in 1795: "First and foremost, the Christian Church decreed that species were created by God and immutable. Second, the men of science also tended to accept the fixity of species because of the success of Linnaeus in classifying species of plants."[32] Anna Seward's analysis of *Zoonomia* in the *Memoirs* follows her father's lead in disapproving of Darwin's irreligious ideals. She appeals to natural theology and the benevolent divisions drawn by the Creator as something that may be plainly observed in the natural world, and adamantly upholds species' fixity. She specifically targets *Zoonomia*'s chapter "Of Instinct," in which Darwin argues that behaviors generally imputed to instinct, such as a bird's song or the construction of a nest, actually result from "observation" and "imitation." Despite his early disclaimer regarding the danger of confusing instinct with reason, Darwin implies that some species are in fact capable of reasoning, and goes so far as to make the leveling declaration: "Go, proud reasoner, and call the worm thy sister!"[33] Darwin's challenge to philosophical divisions between reason and instinct, which Seward saw manifested in his hybrid category of imitation, incites her to critique the concept of mutability among species.

Seward separates instinct from reason and imitation, preserving the distinctness of species' particular behaviors and thereby upholding divisions within the natural order. She insists that "instinct cannot be that lower degree of reason which empowers the animal to *observe,* and by will and choice, to *imitate* the actions, and acquire the arts of his species; since, were it so, imitation would not be confined to his own particular genus, but extend to the actions, the customs, and arts of other animals."[34] In the same vein, she continues,

> If the Creator had indeed given to brutal life that degree of reason, which Dr. Darwin allots to it, when he asserts, that its various orders act from *imitation,* which must be *voluntary,* rather than from *impulse,* which is *resistless,* the resulting mischief of disorder and confusion amongst those classes had outweighed the aggregate good of improvement. It is reasonless, will-less instinct, limited but undeviating, which alone could have preserved, as they were

in the beginning, are now, and ever shall be, the numberless divisions and subdivisions of all merely animal life. As attraction is the planetary curb of the solar system, confining all orbs to their proper spheres, so is instinct the restraint, by which brutes are withheld from incroaching upon the allotted ranges and privileges of their fellow-brutes; from losing their distinct natures in imitation, blending and endless.[35]

Seward displays the anxiety of the taxonomist whose worst nightmare is that which does not fit: species lacking "distinct natures" so that entire orders become unidentifiable, unclassifiable, "blending and endless." Countering this chaos, she invokes a Christian doxology adapted to substantiate the fixity of species ("as they were in the beginning, are now, and ever shall be"). The social injunction to be "content in one's station" rather than "imitating one's betters" echoes clearly.

It is crucial to note that in this passage and throughout her critique of biological constructions in *Zoonomia,* Seward specifically targets Darwin's employment of concepts functioning within the contemporary discourse of literary plagiarism. "Imitation," "improvement," "voluntary," and "instinct" all contained literary connotations to fuel Seward's contentions with this particular chapter of Darwin's zoological study, and suggest that she used the context of zoology indirectly to naturalize her protests against Darwin's plagiarism of her work, establishing comparisons between natural history and literary criticism to advocate poetic, as well as natural, order.

Failed literary imitation, as a form of stylistic plagiarism, constituted a problem that Seward found analogized in Darwin's theory of *zoological* imitation. The threat posed by zoological imitation to the coherence of a "particular genus" correlates with the threat literary imitation poses to coherence of style and voice, risking aesthetic failure. Thus, when Darwin's biological conception of imitation is applied to the poet's "particular gen[i]us," authors, as well as natural species, are in danger of "losing their distinct natures in imitation, blending and endless." To Seward, Darwin's plagiarism of her work in *The Botanic Garden* constitutes an aesthetic failure because the style of the original author (Seward) disrupts that of the imitator (Darwin). The element of improvement is indispensable to an aesthetic judgment of Darwin's imitation, which helps explain Seward's claim that "the resulting mischief of disorder and confusion amongst" poetic styles "had outweighed the aggregate good of improvement." For Seward, Darwin's improvement to her verse is not improvement at all because the incoherence resulting from this mixture of two distinct styles makes his imitation unclassifiable within the poetic order and untenable under the laws of nature. Zoological imitation analogizes failed aesthetic imitation in that both result in degeneration, as opposed to the positive improvement and stylistic coherence found in legitimate cases of imitation.

The illegitimacy of zoological imitation is explained in part by Seward's assertion that imitation requires volition and therefore can occur only in humankind: "*imitation* . . . must be *voluntary*." From a theological standpoint, volition (through reason) makes humans capable of error and improvement, and thus accountable to God for their actions. Seward presses this point, exploiting Darwin's already notorious reputation as an atheist by informing her readers that "to have admitted . . . the unblending natures of instinct and reason, must have involved that responsibility of man to his Creator for his actions in this his state of trial, which Dr. Darwin considered as a gloomy unfounded superstition."[36] Here Seward aligns herself and Darwin with opposing sides of a natural-historical controversy. Whereas Linnaeus was the first to explicitly include human beings within a formal classification of plants and animals, ordering them among monkeys, apes, and sloths, Buffon held more firmly to Cartesian dualism and preserved the distinction between humans and animals.[37]

For Seward, Darwin's erasure of the distinction between reason and instinct also consequently erases the distinction between the categories of "voluntary" and "involuntary." Within the scope of literary criticism, in which voluntary or conscious borrowing represents one of the basic elements of plagiarism, voluntary imitation implies responsibility for resulting aesthetic failings, as well as successes. Seward thus seizes on what she views as Darwin's attempt to elide being both "accountable to God for his conduct," and accountable to principles of literary criticism for consciousness of his plagiarisms. The obviously voluntary nature of Darwin's plagiarism of Seward's poem can be contrasted to an instance in which Seward admits her "unconscious" or "involuntary plagiarisms" of Chatterton in a letter of 1800 (5:273). Because Seward equates "involuntary" with instinct, where instinctual actions can accrue no retribution, her involuntary plagiarisms differ from Darwin's in that she cannot be held accountable, which accords with developing Romantic standards about acceptable imitation. If we can assume Seward's aesthetic success, Chatterton's lines had essentially become inseparable from her poetic identity so that her narrative voice remains whole and intact, whereas in the case of Darwin's imitation, Seward implies that his incorporation of another's poetic style creates hybrid monstrosity.

At the heart of Seward's defense of the preservation of instinct lies her belief in a core poetic identity. Seward presents instinct as "resistless," "limited but undeviating," and as containing individual species "within their proper spheres." In the context of contemporary discussions of plagiarism, her belief in instinct's inherent fixity is consistent with her concern with the poet's unique style or narrative voice: that which ensures that writers, like natural species, "are withheld from incroaching upon the allotted ranges and privileges of their fellow[s]." In his classic study of the critical tradition, M. H. Abrams notes "the tendency in Pope's own lifetime to identify the element" of individual, "natural genius with those instinctive activities of animals, which because they evolve entirely from inherited dispositions, are examples of unlearned behavior *par excellence.*"[38] To exemplify this connection of natural instinct with poetic identity, Abrams quotes Milton, who "had equated Shakespeare . . . with the instinctive singing of a bird" in *L'Allegro,* where Shakespeare is said to "Warble his native wood-notes wilde."[39] This ornithological analogy, equally prevalent in the Romantic era, highlights both biological and literary implications for Seward's indictment.[40] While Darwin discredits "instinctive singing" by arguing that birds observe and imitate an "artificial language," so that "a [turkey] hen teaches this language with equal ease to . . . ducklings," and nightingales "never sing till they are instructed," Seward counters by asking, "Wherefore, since the ear of the feathered warbler is open to the immense variety of strains, poured from the throat of birds of *other plume,* whence its invariable choice of the *family* song?"[41] Like "the feathered warbler," a poet is exposed to the works of other writers but, where there is originality, the instinctive genius adheres to his particular style and keeps his poetic identity intact. This is consistent with Seward's notion that Darwin's borrowing results in a hybrid text of internal incoherence due to his unsuccessful incorporation of another's "song." Through what Seward construes as Darwin's denial of instinct, he undermines poetic identity and thus poets' ability to claim property in their works. Whether Darwin intentionally insinuated literary discourse into his discussion of biological imitation, drawing a witty parallel between the two realms, is unclear, but Seward's singling out of concepts current to literary plagiarism to represent her entire critique of *Zoonomia* clearly indicates her critical agenda whereby natural order confirms poetic order.

SMITH AND THE ORDER OF POETICS

In her efforts to establish the taxonomizing authority behind literary criticism, Seward's main objective is to situate her verse in the order of poetics. Both the *Memoirs* and her letters are filled with persistent attempts to teach readers to recognize her own poetic style. This didactic training includes periodic disavowals of various poems falsely imputed to her, emphasizing characteristics that should have precluded the possibility of her authorship. In one instance, she distances herself from what she considered to be a very poor poem, written by Darwin and to which he signed *her* initials (3:154). And Darwin continued to destabilize her poetic identity even from beyond the grave: following Seward's death, the first poem to gain her nationwide acclaim, her *Elegy on Captain Cook* (1780), was rumored by Richard Lovell Edgeworth to be Darwin's work.[42] Darwin thus trespasses on her not only as a plagiarist, but also (for lack of a better term) as a *reverse* plagiarist, attributing some of his verses to her, and as the supposed author of some of her poetry. Several reviewers of the *Memoirs* admit the justice of Seward's public reclamation of her verses from *The Botanic Garden,* and this acknowledgment is less expected, and her action more courageous, than modern audiences may suspect, for "during the Romantic period, it was extremely rare for a male author to be persuasively charged with plagiarism from a female author."[43] After all, "if men could assimilate her person, then why could they not assimilate her personal expressions as well"?[44] Seward possessed a strong sense of independence—she never married—and her use of criticism to correct not only what she felt to be literary but also sociopolitical, moral, and scientific wrongs, manifests one way of compensating for this vulnerability of women's verse and of "women" more generally. As a critic, she hoped her judgment would guide posterity in configuring the order of poetics, and as a poet, she hoped to find a place within that order. However, efforts to champion women's education and to expand their participation in literature by no means ensured that she endorsed the works of her female peers. She wished works to be appreciated distinct from the author's sex.[45] Therefore, when the sonnets of her contemporary Charlotte Smith threatened both Seward's place in the order of poetics and the order's stability on a grander scale, Seward's reaction was caustic.

As I noted, for many modern critics Seward's attacks on the plagiarisms in Smith's *Elegiac Sonnets* signify jealousy of Smith's popular success. And Smith's literary alliances with William Hayley (by whom

Seward felt rejected) and William Cowper (whom Seward viewed as egotistical and unpatriotic) undoubtedly distanced Smith from Seward's critical favor.[46] However, Seward's literary naturalism provides further explanation, for Smith's plagiarisms enact exactly the sort of disorder among poetic identities against which Seward most strongly protests in her critique of Darwin. According to Seward, Smith's appropriations from other poets disfigure her *Elegiac Sonnets* so that they become monstrosities, "made up of hackneyed scraps of dismality, with which her memory furnished her from our various poets" (2:287). Seward further complains, "I do not find in her sonnets any original ideas, any vigour of thought, any striking imagery—but plagiarism, glaring and perpetual;—whole lines taken verbatim, and without acknowledgment from Shakespeare, Milton, Young, Pope, Gray, Collins, Mason, and Beattie" (2:223–24). How is the literary naturalist correctly to identify and attribute a poem when it contains a "mosaic" of authorial styles?[47]

Seward's accusations against Smith emphasize the kind of chaos and unimprovement she discerns occurring in *Zoonomia,* whereby "the numberless divisions and subdivisions" of poets are not confined to their "separate spheres" and "los[e] their distinct natures in imitation." In contestation of Seward's charges, scholars such as Adela Pinch, Susan Wolfson, John M. Anderson, and Paula Backscheider posit alternative valuations of Smith's plagiarisms, suggesting that when borrowings are written to be recognized and taken from familiar authors, their hybridity becomes something of more productive potential.[48] Smith's appropriations are thereby associated with a larger trend characterized by Robert Macfarlane as occurring in the works of numerous writers of the Romantic era who viewed literary predecessors "as a chorus, a multitude of past voices which added depth and definition to their own poetry."[49] Seward herself confirmed that "imitative traces, of one kind or other, can be found in all works of imagination" (2:183). Yet Seward's complaints against Smith endure because this incorporation of the literary tradition through inclusion of other poets' recognizable styles, of course, increases the risk of incoherence in the authorial voice and, from a critical standpoint, failure to unify the chorus into a single ventriloquization would leave Smith's borrowings unimproved, producing to Seward's ears a cacophony.

A poet's stylistic coherence or incoherence, and the ease with which his or her style can be seamlessly appropriated vitally influenced placement in Seward's order of poetics. She delineated the hierarchizing effect of successful imitation, explaining that "when a great genius condescends to imitate a less, he always excels him; and then the authors, from whom he took, sink, eclipsed, into darkness, if not into total oblivion" (2:183). Interestingly, the obliterating result of "great" poets' stylistic unity lends to the poetic order an air of natural competition for survival, and natural competition is an idea found in Darwin's *Zoonomia*.[50] This struggle for supremacy dramatizes Seward's emphasis on ranking poets and identifying poetic styles; it also indicates that to be a great poet is to be a great imitator, and vice versa.

In the hierarchy of greater and lesser poets that earlier we saw more minutely systematized into classes, Seward wished to designate a definite classification for Smith. Her reaction to a review of Smith's sonnets in the *Gentleman's Magazine* is telling: "Smiled you not to see the reviewer . . . gravely pronouncing, that it is trifling praise for Mrs. Smith's sonnets to pronounce them superior to Shakespeare's and Milton's? O! rare panegyrist!. . . . these hedge-flowers to be preferred, by a critical dictator, to the roses and amaranths of the two first poets the world has produced!!!—It makes one sick" (1:162–63). If she does turn a little green, Seward offers only a derisive smile to what she sees as a complete contravention of poetic order. Seward included Shakespeare and Milton in the long list of poets poorly imitated by Smith and, precisely through insistence on her stylistic hybridity, characterized this rival as a lesser poet, not to be compared, and certainly not "preferred," to "the two first poets the world has produced." More importantly, since Seward's first class of poets is arranged according to "those who are at the head of some particular branch in their science," and the "particular branch" under discussion is the sonnet—a branch in which Seward herself claimed some dominion—praise of Smith was especially galling. As we shall see, Seward endorsed Milton's as the model of the legitimate sonnet, but her placement of Milton at the head of the epic suggests that she intentionally left an opening for supremacy in the sonnet, envisioning herself filling the void. Her relentless fixation on Smith's stylistic hybridity constitutes an effort to label this competitor as a lesser poet, unworthy of heading the sonnet; and, in the context of Seward's order of poetics, Smith's formal choices within the sonnet further secured this relegation.

THE SCIENCE OF SONNETEERING

Seward's literary naturalism reinforced her opposition to Smith in what Seward termed "the sonnet claim." By "claiming" the sonnet, women writers not only located themselves within a masculine tradition, but also threatened to make what Daniel Robinson calls "a bold statement of intellectual and poetic superiority, an implicit act of self-canonization."[51] In the preface to her 1799 collection *Original Sonnets on Various Subjects,* Seward repeatedly refers to, or quotes references to, sonnets as a "species" or "order," and examines a particular "specimen." Her frequent employment of this (Linnaean) taxonomic terminology signals her concern for the classification of, and strict adherence to, this literary genre. She announces her devotion, with only nine exceptions, to the Miltonic model and declares this structure of sonnet alone legitimate. In so doing, Seward takes aim at Smith's own preface to her *Elegiac Sonnets,* in which Smith defends her variances from the sonnet form. Seward's concern is thus for the corruption of the sonnet as a species.

For much of the eighteenth and well into the nineteenth century, natural historians stressed the discovery of new forms.[52] In the *Memoirs,* Seward extols Darwin's *Botanic Garden,* claiming that it "forms a new class in poetry. . . . Nor is it only that this composition takes unbeaten ground, and forms an additional order in the fanes of the Muses, it forms that new order so brilliantly, that though it may have many imitators, it will probably never have an equal in it's [*sic*] particular class."[53] The difference that prompts Seward here to praise Darwin while she condemns Smith lies in the effect of their separate literary experiments on Seward's order of poetics. She emphasizes origins as well as originality. Her notion of species existing "as they were in the beginning, are now, and ever shall be," and as "confin[ed] . . . to their proper spheres," extends to her conception of species of verse. As a "new class," *The Botanic Garden* represents its own legitimate form that, though newly "discovered," fits into its appropriate place of classification, as so many new discoveries of plant species fit within the Linnaean system: filling gaps, making connections more complete, and thus bringing the taxonomic structure closer to a natural order that would reflect divine design.[54]

By designating Darwin's poem as a new class, Seward is repaying a compliment in kind. Darwin had previously credited Seward with a new poetic form, describing her monodies as "Epic Elegies" (5:262). And William Hayley similarly identified the originality of her 1784 *Louisa,* writing, "if your friend Darwin adored you as the inventress of the epic elegy, he ought to renew his adorations to the inventress of the poetical novel."[55] Seward thus perceived herself as participating in this discovery of new forms that fill their legitimate spaces, constituting positive progress in the configuration of poetic order. She avoids aligning her poetic innovations with formal hybridity by instead emphasizing her works' original quality.[56] She sees Smith's sonnets, on the other hand, not as a new discovery of an original form, but as a degenerative imitation of a species already in existence. In the Buffonian lexicon, *degeneration* is a catchword "implying a decline, weakening, and degradation of an original ancestral form."[57] Similarly, Seward underscored the possibility of degeneration in literature by presenting her collection as *Original Sonnets,* belonging to the lineage from the Petrarchan tradition through Milton, and thus antithetical to Smith's self-identifyingly hybrid *Elegiac Sonnets.*

In this vein, Seward additionally differentiates her poetic originality from Smith's borrowings through the illustration on the title page of her *Original Sonnets,* portraying a woman holding a lamp in beams of direct sunlight that suggest autonomous creativity and imagination. Seward's chosen image arguably intentionally contrasts with the initial illustration in Smith's *Elegiac Sonnets* that depicts a woman inspired by the moon's reflected (borrowed) light, incorporating symbolism associated with knowledge of nature through books and through Diana, "Queen of the Silver Bow." Seward thus asserts a strong visual claim, distinguishing her greater originality from Smith's plagiarisms and formal hybridity that she views as degenerating poetry.

Her disgust with Smith's degenerative deviations in form is further elucidated through Seward's treatment of a similar poetic violation in Southey's *Thalaba.* Seward "protest[s] . . . against [Southey's] frequent and licentious change of measure," declaring that "the practice opens a door to much revel-rout, and confusion in poetry, blending its various orders till all distinction amongst them is lost" (6:92). The overlapping vocabularies she applies to zoological and poetic orders here are striking, particularly in her predictions of "confusion," "blending," and loss of "distinction." In contrast, Seward presents the legitimate sonnet as distinct, situated in its designated place and closing gaps within the poetic taxonomy: "It is the intermediate style of poetry, between rhyme and blank verse" (2:226).

In Seward's preface, she defines the legitimate sonnet to obviate the classification of Smith's poems *as* sonnets. According to Seward, Smith's poems constitute "minute Elegies of twelve alternate rhymes, closing with a couplet, which assume the name of Sonnet without any other resemblance to that order of Verse, except their limitation to fourteen lines."[58] Further accentuating the unclassifiability of Smith's

The initial illustration in Charlotte Smith's *Elegiac Sonnets,* 6th ed. (1792), depicting a woman inspired by the moon's reflected (borrowed) light, and incorporating symbolism associated with Diana, "Queen of the Silver Bow." (RB 434679, The Huntington Library, San Marino, California)

Illustration from the title page of Anna Seward's *Original Sonnets*, 2nd ed. (1799), arguably responding to Smith's lunar imagery by depicting a woman holding a lamp in beams of direct sunlight, suggesting more autonomous originality and imagination. (RB 306458, The Huntington Library, San Marino, California)

works, Seward quotes an additional definition of the sonnet printed in the *Gentleman's Magazine* in 1786, stipulating that "Little Elegies, consisting of four stanzas and a couplet, are no more Sonnets than they are Epic Poems."[59] What exactly Smith's "sonnets" are, then, remains undiscovered, unknown—and their lack of distinction (let alone of definition) presents an unwelcome challenge to the order of poetics. In Seward's struggle for supremacy in the sonnet, she forces the question: if Smith's poems are not sonnets, then how can Smith head this "branch [of] science"? To Seward, Smith's formal and stylistic hybridity identified her as a lesser poet, and the widespread popularity and numerous imitations of Smith's poems only revealed "the odd taste of the public," doubtless adding urgency to Seward's didactic efforts and to her faith in a more discerning posterity (2:287).

ANTICIPATING THE NEXT (DE)GENERATION

Although Seward worked to establish the concreteness of poetic order, she herself has become a liminal figure within modern period divisions, and this liminality dramatizes the tensions bound up in Seward's literary naturalism. In her efforts to preserve the order of poetics, and her place within it, Seward enacts conflicts central to contemporary, multilayered discourses of classification through the competing objectives of fixity and dynamism. Her poetic order can appear conservative in its desire for fixity and easily recognizable and classifiable poetic identities, but where she is most seemingly conservative she is also most active in advocating authorial rights through the principles of style and literary property—rights that authors would further negotiate in both courts and literary criticism throughout the Romantic era. Seward's emphasis on the importance of origins and originality participates in the developing ideology of "high" Romanticism, but she complicates this ideology when she regards stylistic and formal hybridity as degeneration. While Wordsworth would also defend the personal property of poetic style, he evaluated formal hybridity differently. By including the category of "composite orders" in his own poetic taxonomy of the 1815 "Essay, Supplementary to the Preface," Wordsworth affirmed his era's increasing comfort with the dynamism that remains strongly associated with the Romantic movement and its imputation of originality to formal hybridity in such works as the *Lyrical Ballads,* Shelley's lyrical dramas, and Smith's *Elegiac Sonnets.*[60] Although hybrid literary forms arguably reinforce even as they elide generic tradition, Seward anxiously predicted their potential for "revel-rout," rather than positive progression, in literature. Her resistance to synthesizing these tensions in her literary naturalism denotes Seward's proximity to, and also her displacement within, the Romantic-period values that have most influenced subsequent canon making.

While fearing readers' inability to evaluate her work properly in relation to that of Charlotte Smith, Seward was not alone in directing critical charges of plagiarism against this poetic competitor.[61] Smith's incorporations of other writers' verse and natural history materials into her poems display a different understanding of plagiarism and originality, offering an alternative framework for science's collaborative possibilities in a literary context. Examining Smith's borrowings and literary engagements with natural history as a means to novelty, the next chapter also uncovers a devastating oversight in John Aikin's recommended mode of poetry with consequences for Smith and the fate of this broader movement of scientific literature.

4

Plagiarism and the Poet-Naturalist

Charlotte Smith's Collective Originality

Anna Seward's allegations of plagiarism against Charlotte Smith, explored in the previous chapter, dramatize the risk of Smith's poetical borrowings throughout her literary career. Although Smith strove to satisfy contemporary standards about plagiarism, she also conceived of literary borrowings as conducive to originality. Indeed, drawing on natural history's collaborative goals and collecting practices, she achieved what I term "collective originality," incorporating other poets' verses while emphasizing her own autogeneity. While Smith used natural history to attain poetic novelty and authority, she later doubted science's ability to fulfill this literary potential. Establishing herself as a poet-naturalist, near the end of her life she made the startling realization that copying nature could draw accusations of plagiarism as easily as copying art. As I will show, this discovery reveals one of the reasons for the inability of contemporary writers to sustain this movement merging science and literature. More immediately, understanding how natural history shaped Smith's perception of originality provides an important rejoinder to accusations against her as a plagiarist, such as those leveled by Seward.

Like Seward, Smith wielded natural history as a weapon of literary criticism, and Seward's attacks on this rival sonneteer's poetical plagiarisms provoked Smith's enmity in return. Several of Smith's letters satirize Seward's character and literary success and, in her most scientific text, *The Natural History of Birds* (1807), Smith rejects both Seward's claim to the title of poet and her charges of unoriginality.[1] Seizing on Seward's critical tactic of ranking writers within her poetic order, Smith wryly classes Seward's literary identity within an alternative, sociobiological taxonomy and differentiates her own conception of authorship.

With her poetic fame, Seward became widely known as the "Swan of Lichfield," a sobriquet in the tradition of Pope, "the sweet swan of Twickenham."[2] Seward prided herself on this ornithological identification, which Smith consciously undermines in her study of birds. Deeming the swan's classificatory order "not so very interesting," Smith jeers that the swan "has been called the emblem of the poets. I know not why, as Wild Swans are gregarious, that is, they assemble in flocks, and the poet is not, I think, a very sociable animal."[3] Her distinction of poetic personality from sociability reflects a Romantic-era shift redefining what constitutes poetic character. Smith alludes to Seward's carefully crafted reputation as a mentor to aspiring poets whose visits and letters were welcomed by the "gregarious" swan, and to her earlier participation in poetic societies such as that in Lichfield and the Batheaston Circle, popular in the 1770s and '80s, within which amateur versifiers "assemble[d] in flocks" and received derision from London critics.[4] Smith thus associates Seward with this earlier movement of sociable poetry focused on inclusive and sentimental virtues of friendship and benevolence, a poetics that the succeeding generation viewed as, in Wordsworth's words, "gaudiness and inane phraseology." Exemplifying Romantic-era mockery of sociable verse, one of Felicia Hemans's "favourite quotations was the satire on the Lichfield coterie [centering on William Hayley and Seward], which she would repeat with exquisite humour: 'Tuneful poet! England's glory,/Mr. Hayley—that is *you.*'/'Ma'am, you carry all before you,/Trust me, Lichfield swan, you do!'"[5] In the late eighteenth and early nineteenth centuries, sociable poetry appeared paradoxically and laughably solipsistic as well as outdated. The swan's affiliation with Pope only accentuates Smith's underlying hint that Seward's sociable verse, like that of Pope and the Augustans, is antiquated. Additionally, scoffing at mythology's assimilation of Leda's swan "with infinite power," Smith clarifies that although "the ancients supposed, that the swan . . . sung most melodiously just before its death," this bird "makes only a sort of snorting noise now and then at particular seasons."[6] By denying Seward divine inspiration and reducing her poetic "song" to a ludicrous "snorting noise now and then," Smith takes a pithy and barely veiled public vengeance on Lichfield's Swan.

Asserting her own prominent place within the poetic hierarchy, Smith contrasts this Sewardian, sociable, and woefully unpoetic swan with her description and versification of the nightingale. According to her, the nightingale "is the most known and admired of all the song-

sters, and is celebrated by the poets more than any other of the feathered race"; further, it is "a solitary bird, and though it really sings all
day, is usually celebrated for it's [*sic*] song during the night; when from
a thorn or low shrub in the hedgerows it is heard to peculiar advantage,
as the rest of the feathered choristers are silent, and the note is sweeter
and more varied than that of any other bird."[7] Even as Smith alludes to
conventional comparisons between the poet and the nightingale, her
emphasis on the bird's solitude and range of originality associates this
feathered songster with contemporary poetry of developing Romanticism. Identifying with the nightingale and breaking from the swan's
sociable poetry, she places herself at the center of a literary tradition
that influences and authenticates this current movement. Smith's poetic
popularity originated in her sorrowfully-themed *Elegiac Sonnets* (1784)
that Seward dubbed "everlasting lamentables"[8]; thus, when Smith explains that "the voice of the Nightingale is considered generally as ex-
pressive of melancholy" and exemplifies the bird's depiction with two
of her own sonnets, followed by Wordsworth's lines on the nightingale
from *Lyrical Ballads,* she aligns herself with this new standard of poetry,
a standard that Wordsworth granted she was instrumental in establishing.[9] Despite Seward's accusations that Smith's poems were "full of notorious plagiarisms, barren of original ideas," Smith was in fact acutely
concerned with literary originality, especially in verse.[10] Considering her
poetry to be her most serious and lasting artistic achievement, she participated in debates over the conventions of her craft and wrote verse
that helped determine her era's changing perception of those conventions while emphasizing the natural sciences.[11]

Modern critics who have addressed Romantic-era women writers'
interactions with natural history gravitate to the works of Charlotte
Smith with good reason.[12] In her verse, Smith oscillates between precision and poetics, between asserting erudite knowledge of species' scientific and common (or vernacular) names, usually providing the missing
appellation in an endnote. Filled with expansive details of the physical
descriptions, the locations, and her personal observations of biological
species, her notes often respond to the most prominent male naturalists
of her day. My study analyzes these scientific engagements primarily in
three of Smith's late texts: *Conversations Introducing Poetry: Chiefly on Subjects
of Natural History* (1804), *The Natural History of Birds* (1807), and *Beachy Head,
Fables, and Other Poems* (1807). Although Smith aligned herself with the
nightingale, I show that her contrasting depictions of the nightingale

and the swan in fact enact an ideological tension existing within her own works. Smith's image of the poet as "solitary," as "not . . . a very sociable animal" participates in a Romantic ideal of autonomy undermined by her notes and poetical borrowings. While she does not practice Seward's brand of sociability, I argue that Smith's poetry produces an originality that is paradoxically collective and thus complexly interrogative of the trope of solitary genius. Aware of increasingly absolute theories of poetic autonomy, she continued to participate in traditions of imitation, borrowing, and improvement, resolving this tension between theory and practice largely by establishing a collaborative notion of literary originality through natural history techniques that blurred forms of authorial identity and evoked networks of both amateur and professional involvement. In this mode of "collective originality," strategically conjoining the collaborative mindset of natural history with literary concepts of individualism, Smith's verse exploits untenable claims to isolation typical of Romantic-era poetic personae, and displays both the possibilities and problems of scientific poetry.

The Poet-Naturalist

When Charlotte Smith added endnotes to her 1786 third edition of *Elegiac Sonnets*—"restor[ing]," as she wrote in her preface, "borrowed" lines and ideas "to the original possessors," including Shakespeare, Milton, Pope, and Gray, in reaction to accusations of plagiarism—she also seized this opportunity to create notes displaying her knowledge of natural history.[13] These notes provide, for instance, the Latin and full common names of the wood anemone and an extended description of the clematis plant. Although literary critics often attribute her use of scientific notes to the influence of Erasmus Darwin's *The Botanic Garden,* her noted edition predates Darwin's *The Loves of the Plants* (1789) by three years.[14] Thus, there seems at least as much reason to allege Smith's influence on Darwin as the other way around. At the same time, Smith's scientific notes contribute to a tradition already in place. As M. M. Mahood explains, "the 'Poet of the Botanists', John Scott of Amwell . . . was in fact only one of a number of poets who began from the 1770s onwards to pack their verses with the names of wild flowers," each accorded its footnoted Latin binomial and, often, expanded commentary, and "this practice fitted in easily with the Augustan poetic; readers expected poets to be informative, as Horace had advocated, and Virgil exemplified in his *Georgics*."[15] While Smith's sex made scientific notes asserting her knowledge and personal observations less expected, this practice became increasingly prominent in her poetry.[16]

Defending herself from accusations of poetical borrowings, Smith sought originality through natural history and in this a more promising initial influence than Darwin is John Aikin.[17] Aikin's *Essay on the Application of Natural History to Poetry* (1777) proposed melding science and verse to combat the "frequent," "invidious," and "discouraging" practice of plagiarism in "Modern Poetry."[18] The problem, according to him, consists in verbatim borrowing of descriptive phrases from poetic predecessors as well as "mistaken," "cursory and general" depictions of natural objects, so that "while the votary of science is continually gratified with new objects opening to his view, the lover of poetry is wearied and disgusted with a perpetual repetition of the same images, clad in almost the same language."[19] For Aikin, the solution is clear: poetic unoriginality "is only to be rectified by accurate and attentive observation, conducted upon somewhat of a scientific plan."[20] Critiquing various examples of natural-historical inaccuracies and "servile imitation[s]" in both ancient and modern verse, he urges poets instead toward subjects of scientific inquiry that remain unexplored, controversial, and thus most capable of producing novelty (such as bird migration, the manner in which young birds practice their songs, and descriptions of exotic nations and their indigenous species). By directing poets to uninvestigated aspects of nature, he unites the prevention of plagiarism to the goal of inspiring "fellow-labourer[s]" in the "interesting researches into British Zoology."[21] For Aikin, the original poet must also be a naturalist, and this view resonated with Charlotte Smith.

Engaging with natural history, Smith highlighted her verses' novel observations. In a note to her poem *Beachy Head* she cites Aikin's disapproval of "how many of our best poets have noticed," that is, overused, "the hum of the Dor Beetle (*Scaraboeus stercorarius*) among the sounds heard by the evening wanderer."[22] She differentiates herself from these poetical plagiarisms in Shakespeare, Milton, Gray, and Collins by exemplifying her own poetic originality, declaring, "I remember only one [other] instance in which the more remarkable, though by no means uncommon noise of the Fern Owl, or Goatsucker, is mentioned," and we learn that this singular reference to the species occurs in an earlier sonnet by none other than Smith herself, seemingly making her the only poet to have recorded this phenomenon. Beyond emphasizing her originality, Smith's note delineates the species scientific name, that of its prey, the adept function of its physical structure, and its associated folklore, setting her apart as a poet-naturalist in the manner encouraged by Aikin, capable of infusing her verse with natural-historical knowledge and ingenuity.

Expanding Aikin's catalogue of poetic errors, Smith asserts her authority in part by correcting others' faulty observations of natural objects. In a note to *Beachy Head,* she revises Shakespeare's description of "the Cuckoo buds as being yellow. He probably meant the numerous Ranunculi, or March marigolds (*Caltha palustris*) . . . *but poets have never been botanists.*"[23] Smith's clearly ironic denial of poets' knowledge of natural history mitigates her critique of this male literary legend while establishing her own botanical expertise. She promotes scientific skill as requisite for writing poetry in her *Natural History of Birds,* stating definitively that "the philosopher and poet should both be naturalists."[24] Smith dubs Thomson "The poet, who perhaps of all that wrote after Milton has most accurately described nature," but redresses his claim that nightingales sing only at night.[25] Although Thomson represented the poet-naturalist par excellence of the first half of the eighteenth century, and Smith admires his "beautiful lines on the birds," she stresses that his ornithological "description does not of course enumerate the varieties of the different species of birds. Of thrushes, for example, there are four or five sorts."[26] She thus relegates Thomson merely to aesthetic excellence, displaying that even this "very correct" poet of nature lacks her level of scientific acumen, suggesting her own supremacy as poet-naturalist.[27] And she does not confine such natural-historical critique to fellow poets.

Pursuing originality, Smith versified natural processes not yet fully understood by naturalists. According to Aikin, subjects such as bird migration invest the poet with unprecedented importance and capability because "the knowledge, indeed, requisite for treating this subject in a masterly manner, would be superior to that of the professed naturalist; since this branch of his researches is yet in its infancy . . the poet should think it incumbent upon him to discover and investigate *new facts,* as well as to frame *new combinations of words.*"[28] Thus, rephrasing extant knowledge is not enough. Aikin's poet-naturalists must become better, more knowledgeable naturalists than the naturalists themselves. Just so, Smith's notes evince her "superior" knowledge and investigations into "*new facts*" enabling her to correct famous naturalists, including Linnaeus and Gilbert White. Her poem "Ode to the Missel Thrush," for instance, seems to have been written specifically for the opportunity to refute White's statement that the thrush ceases to sing "before Midsummer," which Smith argues "is certainly an error," offering proof through personal observation: "now I hear him uttering a more clamorous song, the 8th of July, between the flying showers."[29] In a note regarding the fly orchis plant, she insists that Linnaeus inaccurately "esteemed all those [plants] which resemble insects, as forming only one species, which he terms *Ophrys insectifera,*" citing James Edward Smith's *English Botany* for support.[30] Smith even subjects herself to emendment, admitting, "I was mistaken in supposing [the fern owl] as visible in November."[31] This process of correction and intertextuality marks a common occurrence within natural history works as naturalists revised and expanded colleagues' scientific observations as well as their own past assertions, and Smith's adoption of this collaborative and authoritative stance displays her claim as a naturalist. In the Romantic era, scientific texts generally were considered divorced from an authorial personality and thus easily appropriated, but documentation, citation, and accountability became obsessively characteristic of Smith's engagements with science and poetry in reaction to accusations of poetical plagiarisms.[32]

Indeed, generously citing both poetry and science, Smith's endnotes often interchange natural history and poetical history. For example, her note to a sonnet titled "Insect of Gossamer" quotes the late seventeenth-century naturalist Martin Lister, who studied spiders' ability to "convey themselves" through the air on their floating threads, and she uses poetry to substantiate his scientific remarks, referencing verses by Erasmus Darwin and from Shakespeare's *Romeo and Juliet,* with the latter describing Juliet as a lover who "may bestride the Gossamer/That idles in the wanton Summer air,/And yet not fall—."[33] Equating poetic and scientific authority, Smith allows the natural object to float easily between these two modes of thought.

THE POET-NATURALIST AND THE COLLECTOR

In one of her texts for children, Smith's methodological overlapping between poets and naturalists achieves radical realism. Her first three children's works, *Rural Walks* (1795), *Rambles Farther* (1796), and *Minor Morals* (1798), address natural history and include passages of poetry; but it is in her fourth educational text, *Conversations Introducing Poetry: Chiefly on*

Subjects of Natural History (1804), that she most thoroughly theorizes combining poetry and science. Here, the characters Mrs. Talbot and her children, George and Emily, create a collection of natural history poems as a benevolent alternative to collecting physical specimens of animals, birds, insects, and flowers, so that when the children wish to keep an insect, the green-chafer, that their mother helped them identify, Mrs. Talbot replies, "Instead . . . of contriving the captivity of the chafer, let us address a little poem to it."[34] In a like manner, when George brings home a hedgehog, his mother proposes, "We will try if something cannot be made of it, to encrease [*sic*] our collection of animals as subjects of natural history in verse."[35] And Emily later complains of deficits in her collection, stating, "Mama, I have now several little copies of verses on insects, and some on plants: I have the squirrel too, the dormouse, and the hedgehog, which are beasts, but we have none that tell of birds," to which Mrs. Talbot responds, "I have a bird or two hatching for you."[36] In each case, possession of verses *about* the natural object stands in for, and even becomes equivalent to, possession of the object itself ("I have the squirrel too"). However, this collection of poems interchanging poetical subjects and natural objects is not entirely by Charlotte Smith and, while Dahlia Porter suggests that Smith's inclusion of other poets' works participates in the contemporary fashion for compiling pedagogical collections for young readers, there are more complicated issues of borrowing at stake.[37]

Mrs. Talbot frequently remarks on the difficulty of achieving originality when addressing natural objects versified by so many predecessors. Prior to producing her verses "To the Early Butterfly," for example, Mrs. Talbot tells her son, "It is difficult, George, to say anything that is not mere commonplace on so obvious and hackneyed a subject"; similarly, preceding her poem "The Moth," she disclaims, "Like verses on the butterfly, any attempt on the subject of the moth may perhaps be trite," and, again, "It would . . . be difficult to find anything new to say of that most charming of our feathered musicians [the nightingale]."[38] Smith's struggle for ingenuity in verses about such popular natural objects leads her to commit potential plagiarisms.

Expressing concern about her numerous borrowings, Smith's preface to *Conversations* reveals that five of the work's poems are hers while seven are at least partially taken from other writers. She explains, "I suffered some borrowed and altered pieces to remain, which I should have taken out, had I known that I need not have retained them for want of a sufficient number of original compositions," but "as my trespass on others has not been great, I trust it will be forgiven me."[39] This "borrow[ing]" and "alter[ing]" can be seen, for instance, when Mrs. Talbot acknowledges of her poem "Violets" that it "is not altogether my own. Indeed, some of the lines are entirely taken from a little poem, I believe written by [William] Gifford, and I adapted them to my purpose."[40] Smith's earlier statement that the violet "needs no note, it being . . . in *constant requisition* by the poets" helps explain her use of another's poem on this over-versified subject. Since complete originality appears unattainable, she employs Gifford's work, finding there "kindred sentiments and states of mind."[41]

As discussed in the previous chapter, in the Romantic period, plagiarisms could be divided into two critical categories. Culpable plagiarism, which represented a moral indictment, required that borrowings be simultaneously unacknowledged, unimproved, unfamiliar, and conscious, while the more common charge of poetical plagiarism indicated an aesthetic (not moral) trespass of one or more of these four elements of plagiarism.[42] At this time, when incorporating another poet's verses, the borrower's achieved degree of improvement was largely a subjective critical judgment. Although unsuccessful improvement results in stylistic hybridity or monstrosity, successful improvement seamlessly unifies the two poets' "voices" in poetic ventriloquism and justifies any borrowing. In this way, imitation did not necessarily preclude originality, and since Smith determines that "improvement" to Gifford's poem is possible, she removes stanzas, changes words, and engrafts lines of her own with the goal of asserting something "new."[43] For Smith, the claims of poets, like those of naturalists, are subject to revision, and she readily acknowledges her changes even where she seems unsure of the tenability of her improvements, as when she writes of her adjustments to Cowper's "The Cricket," "tho' it is something like sacrilege to change a word of his, you will see I have made a few alterations."[44] In addition to her efforts to comply with contemporary critical standards of plagiarism, she indicates that her borrowings make good, and even ethical, sense within the context of her collection of natural history poems.

Smith interrogates both the ethics and aesthetics of collecting practices in natural history. Discussing the life stages of a butterfly and attainment of actual specimens for closer inspection, Mrs. Talbot declares that in her youth, "I was soon disgusted with the attempt to kill them. It appeared so cruel, to impale an insect on a pin, and let it flutter for hours and even days in misery, that I could never bear to do it."[45] She expounds that "insects taken for the collections of the curious, must probably have resigned their short lives in some degree of suffering, which nature would not have inflicted."[46] Due to the "suffering" they "inflict," collectors of actual specimens thus become repugnant as their curiosity devolves into "cruel[ty]," desiring the object itself more than the knowledge available through study of the living organism.[47] In fact, Smith asserts the impossibility of understanding an organism from a collector's inanimate specimen, for "the birds, or insects, or quadrupeds, though they may be very well preserved, lose that spirit and brilliancy, which living objects only can possess"; and "their forma or awkward appearances, when stuffed and set on wires, always convey to my mind ideas of the sufferings of the poor birds when they were caught and killed, and the disagreeable operations of emboweling and drying them."[48] Her vivid portrayal of brutal and unfeeling taxidermical practices that put living beings on a trajectory of being "killed," "embowel[ed,] and dr[ied]" emphasizes the moral appeal of collecting poems instead. Smith portrays such distortions of collected objects and of scientific observations in her poem "To the fire-fly of Jamaica, seen in a collection," where she laments that the firefly's "faded form" displayed in the collector's glass case ensures, "Never Naturalist shall . . . see thee . . . with transient gleams to glow" (ll. 36, 39). In contrast to the collector, the naturalist delights in the living habits of organisms. Certainly, a few captives must be made in the name of science, so that she explains of a bird in a sonnet's note, "As I have not seen it dead, I know not to what species it belongs."[49] Still, when an animal, bird, or insect has already been "captured" by another, Smith implies that to then capture it oneself is only cruel. Thus, instead of inflicting the "sufferings" consequent to "impal[ing] an insect on a pin," she states, "I contented myself with copying from . . . collections already made."[50]

Smith's literal use of "copying," of course, refers to drawing the collected specimens, but her pervasive concerns about "copying" the works of other poets, and interchanging natural objects and poetic subjects, recall her "forgive[able]" plagiarisms. By retaining natural objects' vitality, recording their living descriptions and behaviors, the Talbots' poetical collection is antithetical to the collector's case. Indeed, through emphasis on "copying" as a benevolent and life-conserving practice, it would seem that within the poetical collection, the only threat of "suffering" lies in over-versification. There is an equation of actual animal suffering with putting that natural object too often into verse, as if there is something torturous to the specimen in "doing the subject to death." There is no point in writing a poem on a natural-historical subject that has already been captured by other poets, leaving no hope of originality or improvement. In such cases, Smith simply directs the reader to the work of another author, as when she writes of the many poems about the cuckoo, that "none seem to me more simply descriptive than one by [John] Logan, which, as it is so very common, and appears in all collections, I will not insert here. It is inserted in 'Poetry for Children', a compilation with some original pieces by miss Lucy Aikin (Phillips, Bridge-Street)."[51]

However, where improvement is possible, Smith "cop[ies]" the versified subject from another, adding her own observations and alterations to create something new. She thus presents her work as more original than unaltered collecting and as more humane than ignoring the standards of plagiarism or re-versifying a previously captured species to result only in poetic overkill. She produces, in other words, not merely a collection, but something that exemplifies what I term "collective originality." While this phrase intentionally puns on Smith's collection of poetic as natural specimens in *Conversations,* I also intend the connotation of "collective" as something relating to or proceeding from an aggregate of individuals. In her poems and notes, Smith incorporates and references others' poetic and natural-historical assertions, contextually making them her own, and therefore new. She thus conjoins the collaborative mindset of natural history with critical standards of Romantic-era originality. Qualifying Aikin's proscription against copying other poets' verses, Smith's improvements ensure the vibrancy and vitality of both the poetic subject and the natural object. By acknowledging the origins of her borrowings Smith helps to justify her poetic plagiarisms even as these acknowledgments also jeopardize the stylistic seamlessness of her improvements and expose her poems to accusations of hybrid monstrosity.

(Non)hybridities

Smith's conflation of literary subjects and natural objects, inviting readers to view poems as biological organisms, raises the question of whether she conceived of her verse borrowings and formal integrations, in, for instance, her *Elegiac Sonnets* (combining elegy and sonnet forms), as the hybrids Anna Seward perceived them to be. Smith's writings suggest that she viewed her work in more original terms. Her notes and educational texts disparage the contemporary fashion for producing artificial varieties or biological hybrids, especially in plants and birds. In the scientific lexicon, hybrids of different varieties within a species are what we now call intraspecific hybrids, while those between different species within the same genus are usually now called interspecific hybrids, and the offspring of an interspecific cross are frequently sterile—mules, for example. Eighteenth-century naturalists generally understood these sterile productions of true hybrids, or "mules," as monstrous, and denigrated artificial varieties, although the hybrids resulting from these latter crosses are more often fertile.[52] Devaluing such hybrids or variations in flowers, the Linnaean botanist William Withering wrote that "desirable as these changes are to the Florist, they have little weight with the Botanist who considers them as variable accidental circumstances, and, therefore, by no means admissible in the discrimination of the species."[53] Oliver Goldsmith, interpreting Buffon, suggests more substantial taxonomic influence for artificial ornithological varieties: "Pigeon-fanciers, by coupling a male and female of different sorts, can breed them, as they express it, to a feather. From hence we have the various names of Croppers, Carriers, Jacobines, Powters, Runts, and Turbits: all birds that at first might have accidentally varied from the stock-dove; and then, by having these varieties still heightened by food, climate and pairing, different species have been produced."[54] The naturalist John Walker likewise admits breeders' ability to generate hybrids that constitute new species but points to the limits of such productions in that "bird fanciers" and "florist[s] . . . have not been able to produce a specific genus."[55]

Smith herself strongly protests against the bird breeders or bird fanciers who propagate hybrids, "the most extraordinary specimens of the power of art over nature" that are of no use or "real benefit to mankind."[56] She turns the critique into a personal attack, claiming, "What is called a Fancier, whether in flowers or birds, is always a trifling and subordinate character. Such men are only full-grown children, and it is well if their folly be not attended with serious consequences to them." Smith describes one such variety of birds produced to meet the Fanciers' "standard of imaginary perfection": "A Pouter is a bird of which the crop is capable of being so much distended with wind, that the animal appears to be without a head," and she scorns the creators of such monstrosi-

Depiction of an English pouter pigeon in Charles Darwin's *Variation of Animals and Plants under Domestication,* 2 vols. (London: printed for John Murray, 1868), 1:137. Darwin writes of the pouter, "If a bird will not, to use the technical expression, 'play,' the fancier, as I have witnessed, by taking the beak into his mouth, blows him up like a balloon; and the bird, then puffed up with wind and pride, struts about, retaining his magnificent size as long as he can" (138). (RB 604489, The Huntington Library, San Marino, California)

ties, stating, "These [bird-]Fanciers are to Ornithologists, what Flower Fanciers are to Botanists."[57] For Smith, the work of fanciers is inferior to that of naturalists, and their "standard of imaginary perfection" artificially perverts species' forms in nature. Thus, according to her, fanciers' production of varieties such as the Pouter "has excited [only] laughter and contempt."[58]

Her "contempt" for these productions of hybrid living forms indicates that Smith would not have desired her production of poetical forms to be associated with hybridity. Biological hybrids' reputed inability to reproduce prompted Edward Young to make the literary analogy, "an *Original* author is born of himself, is his own progenitor, and will probably propagate a numerous offspring of Imitators, to eternize his glory; while mule-like Imitators, die without issue."[59] Similar to Seward's thinking about her poetical novel and epic elegies, Smith's understanding of biological hybrids as monstrous degenerations of original species suggests that she considered her *Elegiac Sonnets* instead to be original and newly discovered "species of poetry," filling its place in the poetic order. This resonates with contemporary poetical taxonomies in such texts as John Newbery's *The Art of Poetry* (1762) and Hugh Blair's *Lectures on Rhetoric and Belles Lettres* (1783). As has been argued by Jacqueline Labbe, although Newbery and Blair each emphasize "the need to differentiate styles and modes of poetry," they also recognize "that these boundaries are permeable" and necessitate the accommodation of original "species of poetry."[60] The constant need to create new categories in literary and natural taxonomies for these newly discovered forms or species indicates the very incompleteness of those orders. Thus, while the production of hybrids may evoke her "laughter," Smith's literature more seriously highlights the gaps and fissures in contemporary knowledge and constructions of the natural order.

Solving Mysteries through Scientific Collaboration

Teased once by Erasmus Darwin for publishing a poem containing scientific errors, Smith was cautiously aware of shortcomings in her knowledge of natural history, but also recognized that her personal uncertainties reflected the need for greater advances in scientific information more generally.[61] She explains in the preface to *Conversations*, "I fear I have made some mistakes, particularly in regard to the nature of Zoophytes; but the accounts of this branch of natural history in the few books I have, are so confused and incompleat [*sic*], that I could not rectify the errors I suspected."[62] By zoophyte, Smith refers to "that link in the chain which unites the animal and vegetable kingdoms," such as corals.[63] Zoophytes are not hybrids but distinct species, and Smith draws attention to their elusion of reliable scientific account. Throughout her work, she poetically confronts such mysteries and accentuates contrasts between the known and the unknown in nature.

Viewing scientific participation as a means to poetic originality, Smith often emphasizes aspects of nature requiring further study and observation. In so doing, she sometimes exposes science's difficulty in dispelling uncertainties, as in her poem "The Swallow," which explores the much-debated subject of bird migration, and concludes,

> Alas! how little can be known,
> Her sacred veil where Nature draws;
> Let baffled Science humbly own,
> Her mysteries understood alone,
> By *Him* who gives her laws. (ll. 66–70)

Since Smith debuted this poem near the end of her *Natural History of Birds,* her text's previous pages contradict this description of science's "baffled" and "humbl[ed]" state, establishing that much, in fact, "can be known" about Nature and that she herself actively contributes to discovering nature's "laws."[64] Playing on a basic tenet of natural theology, she references God's inscrutable principles paradoxically to imply that these laws' very existence make nature potentially knowable and predictable. By negotiating nature's un/knowability, Smith highlights a tension bound up in the poet-naturalist. Romantic poets frequently claimed inspiration from divine authority or poetic genius, endeavoring to bring reconciliation with the unknown, with not knowing, so that poetry itself derives from life's mysteries and wonders. However, as Adam Smith famously declared, the naturalist, unlike the poet, tenaciously works "to get rid of that Wonder, that uncertainty and anxious curiosity excited by" that which does not fit easily into taxonomies or lacks explanation in nature.[65] Thus, in conjoining the categories of the poet and the naturalist, Charlotte Smith's aesthetic often appears torn between accepting the mystery (especially in her poetry) and striving to solve it (especially in her notes). Yet, her poetic articulations of natural-historical mysteries, I argue, also challenge readers, as she does more directly in her educational prose, to get involved in solving those mysteries and furthering what is known; in other words, the point of revealing which subjects require scientific investigation is to inspire that investigation.

Exceeding Aikin's exhortations to poets to "discover and investigate *new facts*," Smith encourages all readers to natural-historical involvement, stressing science's collectivity. For instance, in *The Natural History of Birds,* she urges military men to record first-hand observations of birds' migratory habits in locations like Gibraltar while traveling in their country's service. Lamenting that "young men in the army are rarely taught to have a taste for natural history, and consider everything of that sort as childish and useless," Smith assures these potential contributors that "this branch of science . . . is neither effeminate nor expensive, but leads to much of the best knowledge, that man in any rank or profession can acquire."[66] Her assurance that the study of natural history is not "effeminate" displays the extent to which women had appropriated this science in public perception by the first decade of the nineteenth century. Arguing for broad inclusivity, she encourages readers of various professions and geographical contexts to help "throw some light on questions" that "cannot be settled but by a course of accurate observations made by persons in different parts of the world."[67] Dependent on new facts and discoveries, the poet-naturalist inspires her audience actively to make those discoveries.

Communication, collaboration, and the widespread involvement of both observant amateurs and "professed naturalist[s]" seemed key to the success of natural history throughout the eighteenth and early nineteenth centuries as naturalists sped to document newly discovered species and new perspectives of species' behaviors to provide an ever-clearer and more complete understanding of the natural order. Just as Smith's encouragements of participation in this science echo Aikin's, his, in turn, renew those of the naturalist Thomas Pennant, who may be said to continue the earlier call of John Ray, and so on. Displaying further collaboration in this field, Gilbert White composed his popular *Natural History of Selborne* (1789) from correspondence with the naturalists Daines Barrington and Pennant. Although Pennant was a steadfast British nationalist Smith calls "the British Pliny," his autobiography boasts of his exchanges with the Swedish Linnaeus and the French Buffon, exhibiting that the communal spirit forged in the search for scientific knowledge could sometimes overcome national prejudices.[68] Linnaeus dispatched his students, known as his "apostles," to the far corners of the globe in order to establish crucial networks of communication re-

garding species in various biogeographical contexts.[69] Joseph Banks
received plant specimens and informative correspondence from more
than 126 individuals worldwide, many of whom were not personally
known to him but represented public enthusiasm for his collecting endeavors.[70] Smith herself exchanged letters with James Edward Smith,
the English botanist and founder of the Linnean Society of London,
and her literary interactions with naturalists include, in addition to Darwin, White, and Linnaeus, also Pennant, William Withering, Thomas
Martyn, Colin Milne, John Lightfoot, and Comte Antoine Francois de
Fourcroy, to name a few whose scientific assertions she addresses in her
educational prose and the notes to her poetry, seeking improved knowledge of nature through joint inquiry. Engaging with poets and naturalists with equal intensity, Smith brings this spirit of scientific collaboration and collectivity into the poetic realm through her verse borrowings.

The Jay and the Nightingale; or, Plagiarism and Original Imitation

The editor of Smith's posthumously published *Natural History of Birds*
lauds Smith's success in "distinguish[ing]" herself as an "original writer"
rather than a "mere compiler."[71] Achieving this originality, in addition to
delineating ornithological species' classifications, physical descriptions,
habitats, prey, behaviors, nest constructions, information about eggs,
songs, usefulness to humanity, and connections to history, mythology,
and poetry, Smith refurbishes four fables, drawing on Jean de la Fontaine, Pilpay (Bidpai), and Aesop, interweaving lessons in natural history and morality.[72] She writes *Natural History* as a continuation of *Conversations* in a series of letters from Mrs. Talbot to her eldest son, Edward,
who is only mentioned and never makes an appearance in the earlier
text. Mrs. Talbot charges Edward and George with explaining to their
younger sister, Emily, any parts of these letters that seem difficult to
understand, suggesting that Smith now targets an older and more informed audience.

Interestingly, in this text, Smith's reworking of la Fontaine's fable
"The Jay in Masquerade" warns against the dangers of plagiarism.[73] She
depicts a vain jay who arrays himself in the dropped plumes of a peacock only to be "ridicule[d]" by "all the folk of the feather." Admonish-
ing her youthful readers to "Be what you are . . . Factitious Art can ne'er
attain/The grace of young Simplicity," Smith moralizes against imitat-

ing affected mannerisms and modish fashions. Her ending lines, however, have a different readership in mind. She closes with the flourish,

> And ye, whose transient fame arises
> From that which others write or say,
> Learn hence, how common sense despises
> The pilf'ring literary Jay (ll. 97–100).

Distancing herself from plagiarism by warning others against it, Smith
places herself among the accusers rather than the accused to imply that
her own borrowings should be understood differently from those of the
"pilf'ring literary Jay." Throughout her *Natural History,* she naturalizes
this fable's analogy between birds and poets, and subtly justifies her
own poetic borrowings through portrayals of imitation and originality
occurring in nature.

Smith investigates the "natural" occurrence of imitation within her
section addressing "the sixth, and most interesting order of birds . . . the
Passeres, which includes all the singing birds."[74] According to Smith,
several species belonging to this order, including the starling, canary,
bullfinch, and reed sparrow, imitate the songs of other birds. She often
confirms natural imitation through personal observation: "I once had
a nest of Bullfinches given me, of which one was reared. . . . she hung
in the same room with a very fine Virginia nightingale, whose song she
soon acquired, and went through the same notes in a lower and softer
tone."[75] This account supports Darwin's argument in his chapter "On
Instinct" in *Zoonomia* that birds' songs may be taught or learned through
imitation.[76] Anna Seward disagreed with this assertion in her *Memoirs
on the Life of Dr. Darwin* (1804), perhaps helping to inspire Smith's judgment of Seward's *Memoirs:* "I never read so very absurd a book."[77] From
Smith's perspective, imitation among birds, and her analogies between
birds and poets, legitimizes some imitations among poet-songsters as
well.

In Smith's ornithological examples, the origins of borrowed songs
remain clear. Although the bullfinch imitates the song of a nightingale, that song remains recognizable as the nightingale's, even when
sung by another. Smith thus demonstrates that borrowings do not obscure a song's origination; nor does the fact of imitation necessarily
damage the beauty of the borrower's song. The bullfinch's "lower and
softer tone" ventriloquizes the nightingale's song in the borrower's own
unique "voice." This designates a seamless appropriation that, in literary

terms, qualifies as successful improvement, as when Coleridge absolves Byron of Wordsworth's criticisms: "W. Wordsworth calls Lord Byron the Mocking Bird of our Parnassian Ornithology; but the Mocking Bird, they say, has a very sweet song of his own, in true Notes proper to himself."[78] If Byron possesses an original song and also sings borrowings in his own "true Notes," he maintains the cohesive spirit of his individuality, and thus Coleridge deems Byron's ventriloquizations to be beautiful and justifiable. Just so, Smith's poetic appropriations, adapted in her own voice and meant to be recognized as of other origin, both fulfill contemporary standards of literary improvement and reflect the mode of borrowing she perceives in nature.

Significantly, Smith exempts the nightingale from her examples of feathered imitators and singles out this bird as the one being imitated in her descriptions of both the bullfinch and the canary.[79] Her representation of the nightingale thus differs from Darwin's in his chapter "On Instinct," with which Smith was certainly familiar.[80] Darwin recounts that nightingales "never sing till they are instructed" which leads him "to suspect that the singing of birds [more generally] . . . is an artificial language," acquired through imitation.[81] Similarly, in Aikin's essay, when he recommends young birds practicing their songs as a new natural-historical subject for poetry, he quotes Pliny on nightingales' imitative rehearsals: "The younger sort mediate and receive lessons for their imitation. The scholar listens with great attention, and repeats."[82] Yet, describing the bird's adulthood, Aikin excerpts Barrington's observations that the nightingale possesses a "superiority" in "tone and variety" to all other birds, and even "sings (if I may so express myself) with superior judgment and taste" that is "excessively brilliant."[83] This attribution of superior "judgment," "taste," and "brillian[ce]" invests the adult nightingale with the appearance of autonomous creative ability within aesthetic terms that transfer directly to the evaluation of poetic expression. Although Aikin and Darwin record the nightingale as imitating its own rather than other species, Smith elides nightingales' imitative capacities and emphasizes instead this aesthetic originality. Since Smith overtly associates her poetic persona with the nightingale, her depiction of the bird as imitated but not itself an imitator implies both the nightingale's and, by association, her own originality.

In *The Natural History of Birds,* Smith exemplifies poetic portrayals of various ornithological species, quoting lines from Shakespeare, Thomson, Gray, Cowper, and so on, but supplies two of her own elegiac sonnets only in regard to the nightingale. Displaying her identification with the bird, Smith calls the nightingale "Sweet Poet" and declares her desire "To sigh, and sing at liberty—like thee!" The nightingale's solitude and "mournful melody" dramatize Smith's conception of poetic character and of her own famously melancholic persona in particular. Thus, while Smith cleverly naturalizes the imitations enacted by birds like the canary and bullfinch as a means of legitimizing certain poetic borrowings, she simultaneously correlates her poetic persona with absolute originality through the nightingale.

According with Smith's notions of originality and legitimate borrowings, her fable of the Jay indicates that plagiarisms become "ridicul[ous]" when an author tries to pass off others' verses as his own without acknowledgment or improvement. Since, in the Romantic era, "a work could be considered implicitly acknowledged or 'avowed' if a 'well-versed' reader could be expected to recognize the original," Smith's citations of borrowings often go beyond critical standards for acknowledgment, demonstrating her desire for her borrowings to be recognized.[84] Indeed, her implicit acknowledgment in borrowing from well-known authors, combined with her efforts toward improving those verses, weakens the case against her as a plagiarist even in her earliest writings. With Smith's further addition of citations in the third edition of her sonnets and subsequent works, she puts the question of adequate acknowledgment to rest altogether, as even Seward grudgingly allows when she writes of Smith's *Elegiac Sonnets,* "I have not seen the [third] edition, but am told that she has in that put the quotation marks so disingenuously withheld in the first publication."[85] In her verse, Smith draws on the poetic tradition, determining which sentiments are worth repeating, and this process correlates with her selective and corrective adjustments to naturalists' claims. Through these improvements and collaborations, she transfers to poetry the unceasing progress and newness of thought that Aikin locates in science. Incorporating the assertions of other poets and naturalists, she achieves collective originality by cohesively ventriloquizing those borrowings within her own voice and novel observations.

Still, if Smith sought to dissociate herself from obviously unimproved plagiarisms like that dramatized by the fabled jay, she also felt compelled to suppress that fable from the audience that most needed convincing. Of her poetic fables written for *The Natural History of Birds,* Smith excludes only "The Jay in Masquerade" and one other poem from reappearance in her better-known posthumous publication, *Beachy Head, Fables, and Other Poems,* which targeted a broader adult readership.[86] Since in *Beachy Head*'s "Notes to the Fables" Smith expresses fears of accusations of plagiarism in regard to another of her fables, her omission of the jay poem likely signals her suspicion that, rather than exonerating her, the "pilf'ring literary Jay" could provoke perilous comparisons between herself and *this* bird (instead of the nightingale) and prompt critics to advise that she take her own counsel against illegitimate plagiarisms. Thus, she omits the fable of the jay from *Beachy Head* and finds herself in a precarious position.

THE LARK; OR, THE PROBLEM OF ORIGINAL NATURE

In "Notes to the Fables," Smith admits the fable genre's inherent unoriginality, as stories that are old and frequently retold, yet claims "a degree of novelty" in her contributions of natural history.[87] She thus counters Aikin's insistence that fables' familiarity precludes the introduction of "minute or uncommon relations in natural history," making them "unfit for the display of that novelty which natural history affords."[88] Interestingly, the novelty of one of Smith's fables is indeed compromised, but not in the way Aikin predicted. Assuring her readers that "there is nothing I am more desirous of avoiding, even in a trifle like this, than the charge of plagiarism," Smith explains, "I must in the present instance defend myself." She expresses surprise at finding in James Grahame's *The Birds of Scotland, with other Poems* (1806) "what, if my fables had been first published, I might perhaps have thought very like an imitation." Examining lines from her fable "The Lark's Nest," and similar lines from Grahame, she strikingly posits, "The extreme resemblance of these passages may be accounted for . . . by the observation very justly made that *natural objects being equally visible to all, it is very probable that descriptions of such objects will be often alike*" (emphasis mine).[89] She thus undercuts Aikin's confident assertion that "such is the variety of nature, that original [poet-naturalists], even of the same subject, need not be apprehensive of falling into an uninteresting sameness."[90] Whereas Aikin insists that accurate observations of nature provide inexhaustible multiplicities of thought and expression, Smith now qualifies such enthusiasm and questions the originality available through natural history poetry. Smith understands that species' constancy (in behavior, physical structure, and so on), which enables direct observations "of the same subject" to be repeated and generalized, makes science possible. However, she also now recognizes that this constancy, making the naturalist consistent and accurate, could prove devastating for the poet whose close observations of nature risk appearing as poetical borrowings from other poets closely versifying the same natural phenomenon. Under such circumstances, even relatively new scientific observations and discoveries may not retain their novelty for long, and it therefore becomes a race of who can poetically inscribe the data before it becomes exhausted. To achieve the originality promised by Aikin, poets must indeed continuously "discover and investigate *new facts*," and participate in the verification and debate of scientific assertions, and, for Smith, this quest ends in a disheartening poetic discovery.

In her initial publishing of "The Lark's Nest" in *The Natural History of Birds,* Smith explains that in refashioning Aesop's fable she has "dressed it with a few botanical ornaments, which I think you will allow to be an improvement."[91] She thus adheres to contemporary critical standards, refuting plagiarism by claiming "improvement" to Aesop's work. However, the preemptive defense against Grahame she considers necessary when republishing the fable in *Beachy Head* illustrates that Smith has learned a great irony: the copying of *nature* may lead as readily to charges of plagiarism as the copying of *art*. Where species and natural phenomena are not subject to alterations, "descriptions of such objects will be often alike." Since *Beachy Head* was published posthumously and marks the last volume of poetry she produced, it is impossible to say whether this realization would have discouraged Smith's future use of natural history in her poetry. Her delight in pointing out mysteries in nature requiring further investigation indicates belief that much potential remained for nature to lend novelty to verse. Yet, as I show in my remaining chapters and conclusion, had she lived to see it, an imminent shift in the practice and portrayal of both natural history and originality, and in their accessibility to women, may have had more dire implications for her poetic uses of the science.

"A Solitary Bird"

Smith's (self-)representation of the nightingale embodies the now-preeminent formulation of the Romantic poet as genius who sings in inspired and often melancholy isolation. Her depiction participates in similar self-portrayals propounded by other writers of the era conflating

Romantic authorship with autogenous originality as when Wordsworth claimed that the writer should "owe nothing but to nature and his own genius."[92] Natural history's reliance on collaboration and competitive exchange thus may seem to oppose the era's prevailing aesthetic. However, by presenting herself as a poet-naturalist, Smith bridges scientific and literary expectations. Notes to her poetry create a context for documenting her scientific knowledge, discoveries, and engagements with fellow naturalists, as well as the collective exchange occurring in her numerous borrowings and improvements within the poetic tradition. Smith's elaborately notated interactions with both natural history and poetical history subtly deny the solitary posture of the visionary poet even as she employs that posture to consolidate her claim to originality. Scholars have long recognized the trope of the autonomous poet as "an aesthetic fantasy," a "bogeyman" of the "Romantic cult of the individual genius" that belies a more richly collaborative mode in reality pervading the writings of Romantic-era poets.[93] The period's authors' extensive appropriations of lines and ideas from various sources led Thomas Peacock (of apt ornithological affiliation) to complain in *The Four Ages of Poetry* (1820) that "Southey wades through ponderous volumes of travels and old chronicles . . . and when he has a common-place book full of monstrosities, strings them into an epic," and Wordsworth, Scott, Byron, and Coleridge each likewise "picks up," "digs up," "cruizes for," and "compound[s]" "heterogeneous congeries of unamalgamating manners."[94] Peacock's satirical description of "disjointed relics of tradition and fragments of second-hand observation" suggests his critical assessment of unsuccessful assimilation and enlivens our perspective of these poets' intertextual productions as collections of sorts. While Smith's own immediate, personal observations of nature lend an intriguing corrective to the "second-hand" appropriations Peacock denounces in these male contemporaries, her obsessive "restor[ations]" of verse borrowings and citations of naturalists underscore the Romantic paradox of collective originality.

In the notes to her poetry, Smith's engagements with science's crucial networks of transnational communication arguably gesture toward natural history's cosmopolitan potential. However, science concurrently brings out national tensions as well as opportunities for cross-cultural collaboration. Exemplifying these coexisting possibilities for individual and collective identity, in the first stanza of *Beachy Head* Smith cites geological theories of continental shifts and envisions the politically charged moment of England's isolation, of becoming "divided" from France in

> the strange and awful hour
> Of vast concussion; when the Omniscient
> Stretch'd forth his arm, and rent the solid hills,
> Bidding the impetuous main flood rush between
> The rifted shores, and from the continent
> Eternally divided this green isle. (ll. 5–10)

Characteristically questioning naturalists' assertions, Smith's note clarifies that her own investigations find nothing to support such claims: "I confess I never could trace the resemblance between the two countries." For Smith and her contemporaries, the American and French revolutions and the Napoleonic Wars created political conflict in which natural history was unavoidably embroiled, even as science was often celebrated for its supposed immunity from national prejudices. In this vein, the next chapter explores Helen Maria Williams's revolutionary convictions and attention to naturalists' (particularly geologists') concepts of cosmopolitanism in both society and nature. Investigating science's globalization in the midst of national and imperial strife, Williams asserts influence and originality surprisingly through translations of natural history texts.

Revolution and Geological Sciences

Translations, Beginnings, and Endings

5

Translating Cosmopolitanism

REVOLUTION IN HELEN MARIA WILLIAMS'S
GEOPOLITICAL NATURE

Throughout the second half of the eighteenth century, natural history remained in the public domain, inciting enthusiasm and participation from amateur naturalists as an "open and egalitarian" form of natural inquiry; indeed, contemporaries of the French Revolution sometimes represented natural history as science's democratic ideal.[1] An early proponent of liberal values, Helen Maria Williams hosted renowned salons in London in the 1780s, and later in Paris, drawing visitors of shared political views from all over the world, including many naturalists. Now best known for her eight volumes collectively called *Letters from France* (1790–96), she supplied first-hand accounts to Britons eager for news of the revolution's developments and degeneracies. Modern scholars frequently remark that after 1791 Williams never returned to Britain, residing thereafter primarily in France, and her bold entry as a woman in the political sphere as well as her continued support for revolutionary principles provoked censure from British contemporaries. In this chapter, I expand this critical narrative to demonstrate how her shifting portrayals of France's political history coalesced with her understanding of originality and natural history, as exhibited through her scientific poetry and translations.

Possessing knowledge of the natural sciences and particularly of geology, Williams translated from French into English a number of naturalist works, including J. H. Bernardin de Saint-Pierre's popular novel *Paul and Virginia*, Louis-François Ramond de Carbonnière's essay on Alpine glaciers, and two of Alexander von Humboldt's major works of exploration in Latin and South America. In these translations, she exemplifies a collaborative approach that accords with revolutionary convictions of social inclusion, adding her own political observations, poetic augmentations, and scientific insights. Williams's alterations to these naturalists' works conjure up Romantic-era debates about the extent to which translators may appropriate texts through original interventions. Critical standards for translating scientific works typically required "render[ing] content and ideas as accurately as possible."[2] In each of her translations, Williams mediates dissemination of the original author's ideas, strategically altering their presentation to suit her voice and vision, as well as her personal and political ends. In this mode, a work's origin becomes shared, hybrid, blurred, emphasizing "the inevitable tensions of identity and alterity in all translation."[3] Interestingly, this tension between collaboration and individual identity also "translates" to Williams's egalitarian politics, combining "competitive individualism and representative democracy," and to her representations of natural history.[4]

Natural history's relation to society altered during Williams's literary career, and I argue that she reflects those changes in her translations, and in her respective poetic depictions of Captain James Cook, and of the naturalists Erasmus Darwin and Humboldt. For her, each of these men emblematizes different stages in France's political environment, as well as different forms of cosmopolitanism.[5] Williams's portrayal of Cook anticipates the universal equality promised by sociopolitical revolution that she further epitomizes in her writings on the Swiss Alps, associated with Darwin and concepts of geological revolution. Humboldt, on the other hand, later represents a different kind of cosmopolitanism that marks distinctions between particulars in establishing universal claims. In both cases, the scientific tug-of-war between individual identity and its erasure through universals surprisingly reenacts the translator's struggle between original expression and "invisibility" through textual fidelity. Like many of her contemporaries, Williams subtly correlates Humboldt's ambitions as a naturalist with Napoleonic conquest and individualism as well as with the liberal principles she earlier invested in Cook's scientific, humanitarian cause. Humboldt represents a transitional figure in the professionalization of the natural sciences. For Williams, France's shifting political circumstances compare with natural history's trend toward specialization, moving away from the inclusivity earlier intrinsic to encouraging serious public participation, especially from women, as exemplified by the Cook voyages.

Cook's Cosmopolitanism: "The Friend of Human Race"

Captain James Cook's first voyage to Tahiti, New Zealand, and Australia (1768–71) revolutionized natural history, becoming synonymous with science's collecting practices, as well as with the success of collectively national and international scientific goals. He returned to Britain with an unprecedented number of plant and animal specimens, "as well as charts, journals and calculations that put parts of the world on European maps for the first time."[6] Joseph Banks, the English botanist and later president of the Royal Society, accompanied Cook on that initial expedition and, with his employees, collected one thousand zoological and thirty thousand botanical specimens, increasing the number of known plant species by 25 percent.[7] Banks later promoted these natural historical discoveries to solicit further aid in collecting specimens and information, penning more than twenty thousand letters to correspondents around the globe.[8] Although Cook had secret orders to claim land for expansion of the British Empire, his voyages' explicit pursuit of scientific knowledge appeared to depart from previous conquest-driven explorations.[9] As opposed to the embarrassing losses in colonial America and the imperial brutalities inflicted in India and the Caribbean, Britons took pride in Cook's peacefully motivated discoveries as benefiting all of humanity. His death in the Hawaiian Islands on February 14, 1779, during his third expedition, thus shocked and saddened British readers when news finally reached London on January 10, 1780. Reports of the great voyager's death generally depicted Cook as heroically altruistic to the end. After Hawaiian islanders stole a large cutter, his ship's biggest and most useful boat, he attempted to recover it. According to eyewitnesses a skirmish ensued and, during Cook's efforts to make his men cease firing, islanders overtook and killed the navigator so that he fell victim to his "excessive humanity."[10]

Women writers took great interest in Cook's explorations and in the man himself. Shortly before closing his biographical *Life of Captain James Cook* (1788), Andrew Kippis finds it "somewhat remarkable that female poets have hitherto been the chief celebrators of Captain Cook in this country" and urges that a subject so steeped in science and discovery "may hereafter call forth the genius of some poet of the stronger sex."[11] By the time Kippis published his biography, Cook had in fact appeared in poems by Cowper ("Charity") and Hayley (*Otaheite: A Poem*), but more popular were the verse tributes of Hannah More ("The Black Slave Trade") and especially Anna Seward (*Elegy on Cook*). Despite his

anxieties about Cook's appeal for women, Kippis introduced yet another female poet's "wreath to the memory of our navigator" by including Helen Maria Williams's poem "The Morai: An Ode" as an appendix
to his work.[12]

Active in political, scientific, and literary circles, Kippis ministered
a dissenting church in London attended by Williams, her mother, and
sisters, and he mentored Williams, whose father died when she was an
infant. Kippis ushered into public notice earlier poems by Williams as
well, writing the advertisements for her first three poetic works, which
she published anonymously in the early 1780s.[13] Thus encouraged in
her talent for poetry, Williams also garnered the support of Elizabeth
Montagu, who headed the legendary Bluestocking circle and to whom
Williams dedicated her 1784 poem *Peru*. In 1786 Williams published a
two-volume edition of poems in her own name and already possessed a
wide and adoring audience as a poet of sensibility, famously commemorated in Wordsworth's first published poem, "On Seeing Miss Helen
Maria Williams Weep at a Tale of Distress" (1787).[14] Yet, the sensibility
of this "plaintive Muse," as Kippis calls her, doubtless contributes to his
concern for the emasculating potential of female poets' interest in Cook
and his voyages.

The chief poetic celebrator of Cook, Anna Seward, similarly excelled
in the rhetoric of sensibility; the two poets became intimate friends,
and Seward wrote a sonnet calling Williams her "Poetic Sister." Seward
further complimented the younger poet's verses on Cook, writing in a
letter to Williams, "I have read your glowing poem . . . and felt at once
thrilled and warmed by its solemn fire."[15] In the same letter, Seward
gestures toward her own *Elegy on Captain Cook* (1780), deriding Kippis's
masculinist anxieties while adopting his imposed solidarity of "female poets." She confides to Williams, "I smile to see how curiously he
guards against either you or me growing too vain on the subject of our
poems on Cook,—deploring, as he does, that our hero had no *abler* pan-
egyrists." In a later letter to Reverend Berwick, Seward displays more
bitterness about the backhanded "praise [Kippis] deigns to bestow on
the Muses."[16] Dismissing Kippis's biography of Cook as superfluous
to a public already familiar with the navigator's life, she inverts Kippis's gender and literary hierarchies by deeming Williams's appended
poem as the most impressive part of Kippis's text; for Seward, Williams's "Ode seems the gem of the Doctor's work. It is very sublime.
That young lady's talents are indeed an honour to our sex." Extolling
Williams, Seward thus asserts women poets' ability to illuminate "vivid

and original" aspects of a well-known scientific narrative, while portraying Kippis as failing in the attempt. Paradoxically, Williams's originality owes in part to her poem's reiteration of key images from Seward's *Elegy*, with differences that indicate these female poets' separate agendas, readily displayed through comparison.

While Williams's poem functions, literally, as an addendum to Cook's life, Seward's *Elegy*, published in June 1780, a few months after Britain received news of Cook's death, acts as a biography in itself, relating exciting episodes from his expeditions through "The scorch'd Equator, and th'Antarctic wave" (l. 30).[17] Seward's numerous footnotes, often quoting the published account of Cook's second voyage, highlight resulting discoveries in natural history, reference Linnaeus, and describe new and exotic species, including the kangaroo, giant bat, and "coral rocks" created by "sea-insects."[18] Stressing the voyages' commercial benefits, Seward explains that a plant found in New Zealand contains fibers that "are longer and stronger than our hemp and flax; and some, manufactured in London, is as white and glossy as fine silk. This valuable vegetable will probably grow in our climate."[19] Indeed, for Seward, this importation of economically viable discoveries makes "imperial London" the cosmopolitan center of the world (l. 17). Cook's explorations materially augment the global bounty already enjoyed by Britons in refreshing "cups of summer-ice," "the incense of Sabaean vales" from the "Orient," and Italian silks and "artful song" (ll. 19–23). However, Seward subordinates this celebration of empire's commercial benefits to Cook's conferrals of British charity.

According to Seward, the guiding power of "BENEVOLENCE" motivates Cook's expeditions that convey "garden-seeds," implements, and livestock to South Pacific islanders (ll. 121–32). He plants the seeds both of vegetables and of Christian values, and his charitable intentions starkly contrast with those of "half the warring world [who] . . . dye the distant waves in human gore" in the American Revolution (ll. 208–9). Cook pursues peace even among the cannibals of New Zealand, where "the frowning natives. . . . scowl with savage thirst of human blood!" (ll. 109–12). In Seward's account, most islanders gratefully "Rever'd the stranger-guest, and smiling strove/To sooth his stay with hospitable love," especially the people of "Otaheite" or Tahiti (ll. 115–16). Her description of "hospitable love" cools accounts of Tahitians from Cook's first voyage that fired British imaginations with tales of edenic free-love, a temptation from which Cook abstained (ll. 115–16). His moralizing example to both the Tahitians and his crew uniquely qualify him, in Seward's portrayal, as a hero of sensibility. When part of an iceberg breaks perilously close to Cook's ship in the second voyage, for instance, she records an effusive emotional reaction as he "checks the rising sigh,/And turns on his firm band a glist'ning eye" while care and compassion for his crew "starts the impassion'd tear" (ll. 91–92, 95). Such sentimentalized depictions allow women writers like Seward to align Cook's achievements with virtues that, in Kippis's view, feminize the hero.

Promoting feminine virtues as laudable British values, Seward justifies women's involvement in national and scientific concerns, and represents women as guardians of national principles by appropriating for them the task of mourning and memorializing Cook's death. First imagining the Tahitians' "Morai" or "funeral altar" for Cook, Seward notes that a frantic and bloody female, their "chief mourner wanders around it in a state of apparent distraction, shrieking furiously, and striking at intervals a shark's tooth into her head. All people fly her, as she aims at wounding not only herself, but others."[20] Seward contrasts this violent and foreign female grief with a less frenetic British response, exemplified by "a softer form," Cook's widow, gazing out on the sea from "aloft on Albion's rocky steep," and waiting "in vain" for her husband's return (ll. 247, 249, 251). This more reserved feminine reaction expresses a devotion and dignified depth of loss by which Seward depicts British women as modeling for the Tahitians how to mourn Cook's death, just as Cook modeled morality for the islanders in life. Cook's achievements thus become synonymous with British greatness and with the moral authority of that country's grieving women, implicitly women writers like Seward who, in representation of the nation itself, ensure that "his fame shall rise,/In endless incense to the smiling skies" (ll. 269–70). When Seward asserts that her poem is "In deep accordance to a Nation's woe," she becomes, as Harriet Guest persuasively argues, "the personification of national identity," prompting other poets to hail her as "Our British Muse" and "th'immortal MUSE of Britain."[21] In light of Seward's concerns with gender and nationalism, it is significant that she originally termed the force motivating Cook's voyages as "HUMANITY," and only later changed this to "BENEVOLENCE," likely in reaction to the backlash against the French Revolution. Charged with implications for European class systems, the two words differ in the relationship they suggest between Cook and the people he encountered. Whereas "benevolence" arguably denotes an unequal power relationship that places the European explorer in a position of charitable superiority to the islanders, "humanity" denotes a more equal exchange—it is a leveling term that Seward prudently discarded, and that better applies to Helen Maria Williams's poem on Cook.

Williams embraces a broader humanity than that which Seward emphasized in feminine values and British national identity and, as Williams's funereal title suggests, "The Morai" picks up where Seward's poem ends. While Seward primarily commemorated Cook's life, Williams takes his death as the overwhelming subject of her verse, a subject that lends itself to expressions of both sensibility and universalism. She joltingly interrupts the exotic description of Tahiti that begins the poem to ask, "Whence arose that shriek of pain?" as a means of entering a general meditation on death (l. 27).[22] Refusing to narrow her scope to Cook's particular demise, Williams articulates grief and loss as transcultural proof of our common humanity:

> from the shore where Ganges rolls
> His waves beneath the torrid ray,
> To earth's chill verge, where o'er the poles
> Falls the last beam of ling'ring day,
> For ever sacred are the dead! (ll. 53–57)

Regardless of race or nation, respect for the dead and mourning practices are bound up in death's leveling force, erasing differences. Williams depicts Tahitian funeral rites as similar to those in Britain, and makes the frantic female Tahitian mourner, repeated from Seward's poem, seem more genuinely sympathetic and representative of grief's universality.

When, nearly three-fourths of the way through the poem, Williams finally alludes to the specific death of Cook, her sympathetically established universals strategically frame the man she calls "the friend of human race" (l. 142). Interestingly anticipating Mary Louise Pratt's theorization of "the contact zone," conceptualized "in terms of copresence, interaction, interlocking understandings and practices," Williams designates the space of Cook's encounters with islanders as the "connecting zone," while demonstrating less skepticism than Pratt in a narrative of anti-conquest that "connects" humanity (l. 47).[23] Williams's attention to human similarities in the midst of difference emphasizes "philanthropy" as bi-directional within the connecting zone, extending both ways, as Kippis's biography makes clear through Cook's dependence on the willingness of natives to exchange provisions and goodwill. Williams distinguishes Cook's Pacific encounters from both the brutality of Spanish conquest in the New World and, more audaciously, Britain's contemporary enslavement of Africans (ll. 149–58). In an obvious departure from Seward's nationalism, Williams critiques Britain's violations of the connecting zone as violations of humanity, and separates Cook from a narrowly British identity by designating him instead as "the friend of human race."

Written on the verge of the French Revolution, Williams's poem promotes inclusive, democratic ideas based on Cook's association with natural history.[24] His voyages epitomize science's power to overcome political differences as Kippis records in his biography that, although at war with Britain in the American Revolution, the French Louis XV proclaimed that "such discoveries being of general utility to all nations, it is the king's pleasure that Captain Cook shall be treated as a commander of a neutral and allied power," and scientific communities in America and Spain urged similar sanctions.[25] Cook thereby symbolized natural history's ability to unite nations in the common goals of knowledge and progress, a transnational capacity that Williams also attributed to poetry and stresses in her poem on the explorer.[26]

For Williams, as for Seward, Cook's bereaved widow emblematizes Britain's immortalization of his achievements, yet Williams's poem stresses that this immortalization additionally occurs in all nations of the world, for "natives of the earth / Shall oft repeat thy honour'd name / While infants catch the frequent sound, / And learn to lisp the oral tale" (ll. 195–98). More radically in death than in life, Cook becomes a unifying force; his "tale" becomes another common strain in humanity as an international bond preserved by "the muse of history," a history that is both forward-looking and pervasive of national boundaries (l. 192). And Cook's role in this progressive political history resonates with his reputation's inseparability from progress in the egalitarian sciences of natural history. Williams's 1788 poem thus displays her enthusiasm for revolutionary ideals and exemplifies Cook as representative of republican and humanitarian goals.

TRANSLATING BERNARDIN: "I NO MORE MY LONG-LOST HOME SHALL HAIL"

For Williams, the revolution in France poised Paris to achieve the ideal harmony conjoining natural history, literature, and politics envisioned in her poem on Cook. Upon arrival in Paris on July 13, 1790, she became the self-appointed historian of the French Revolution and, indeed, in her first volume of *Letters from France,* she depicts the revolution as embodying universal humanity, declaring, "Oh, no! this was not a time in which the distinctions of country were remembered. It was the triumph of human kind; it was man asserting the noblest privilege of his nature; and it required but the common feelings of humanity, to become in that moment a citizen of the world."[27] In Williams's portrayal, France's politicians and naturalists strove toward the same goals of bringing order to chaos so that the seeming disarray of the National Assembly's debates indicate a developing political order reflective of that found in nature: "The new constitution arises, like the beauty and order of nature, from the confusion of mingled elements!" (1:1:44). Science and politics were of course traditionally male arenas of discourse, and so, for Williams, a sign of France's progress lies in the extent of knowledge made available to women. In addition to meetings of the National Assembly, women could attend lectures at the Lycée, which was formed in 1785 and frequented "not only by men of letters, but by the most fashionable persons of both sexes," where one could hear lessons delivered "by the most celebrated professors of Paris, on natural philosophy, chemistry, natural history, botany, history, and belles letters" (1:2:130). Lauding women's participation in scientific discussions at the Lycée, Williams remarks, "I regret we have no such institution in London" (1:2:132).

Promoting the revolution as conducive to gender equality, Williams also acknowledges that many British critics disapprove of her political participation and "prophesy that I shall return to my own country a fierce republican" (1:1:66). She attempts to mitigate accusations of gender impropriety by portraying her political involvement as femininely motivated, assuring, "my political creed is entirely an affair of the heart; for I have not been so absurd as to consult my head upon matters of which it is so incapable of judging" (1:1:66).[28] Williams employs conventions of modesty, denying intellectual involvement in politics to defuse critique of her demonstrations to the contrary. Sanctioning instead women's participation through "superior sensibility," she declares that the revolution makes imminent a time "when the human mind has made as many important discoveries in morality as in science" (2:1:213; 1:1:65).

However, as Robespierre came to power and France's sociopolitical virtues turned into violence, Williams's vision of the revolution conjured up not Cook's scientific legacy of universal humanity, but the most horrific encounters of his explorations. In October 1793, while Williams conversed over tea with the author and naturalist J. H. Bernardin de

Saint-Pierre, word arrived that all Britons in Paris were to be arrested and their property confiscated in reaction to successive French military defeats (2:1:6). Shortly thereafter, Williams, her mother, and sisters were imprisoned for two months, first in Luxembourg and then in the convent of Les Anglaises.[29] In prison, the specter of Cook haunted Williams as her friends in the moderate Girondist political party, representing to her mind the ideals of equality and liberty, faced the guillotine at the hands of Jacobin officials, whom Williams called "cannibals" (2:2:149). The military commandant who controlled her prison likewise appeared to her as a figure from the voyages, possessing a "fierceness" that "seemed to be of that kind which belongs to a cannibal of New Zealand; and he looked not merely as if he longed to plunge his saber in our bosoms, but to drink a libation of our blood" (2:1:29). For Williams, even the human sacrifices in "barbarous countries" reported from Cook's expeditions now seem less barbarous than the atrocities in France, where the goal of universal humanity has become a mockery, productive "not of an equality of happiness, but of an equality of misery, throughout the republic" (2:1:26–7, 2:2:166).

During her imprisonment and preoccupation with Cook's voyages of natural history following conversation with Bernardin, Williams translated the latter's novel *Paul and Virginia*, "to cheat the days of captivity of their weary length" and escape her "own gloomy reflections."[30] A botanist and student of Rousseau, Bernardin set his sentimental novel on the island of Mauritius, off the southeast African coast, where he earlier documented indigenous plants. His novel relates how two "children of Nature," raised in the style of Rousseau's pedagogical philosophy, grow up, fall in love, and end in tragedy due to intrudence of European manners and artifice.[31] Williams's preface resituates the novel, originally published in 1788, now "amidst the horrors of Robespierre's tyranny."[32] She explains that part of her translation had been confiscated and "sent to the Municipality of Paris, in order to be examined as English papers . . . and are not likely to be restored to my possession."[33] "Hop[ing] to deserve the humble merit of not having deformed the beauty of the original," Williams alters Bernardin's text in two main respects: by "omitting several pages of. . . . long philosophical reflections" and by interspersing eight of her own sonnets throughout the work.[34]

Interpolating within her translation the eight sonnets, "To Love," "To Disappointment," "To Simplicity," "To the Strawberry," "To the Curlew," "To the Torrid Zone," "To the Calbassia-Tree," and "To the

White-Bird of the Tropic," Williams attributes these poems to the character Madame de la Tour, Virginia's mother. However, these frequent interruptions of the narrative call attention to Williams's presence as both translator and original contributor.[35] Her sonnets reference climatology, ornithology, and medicinal botany. Their content also reminds of Williams's imprisonment, alerting British readers to recognize her poetic persona of sensibility now in a foreign land and longing for home. For instance, in "To the Strawberry," the speaker's sight of this "Plant of my native soil!" invokes lost "scenes of childhood" and, in "To the White-Bird of the Tropics," she ends a poetic meditation on the mysteries of bird migration with the lament, "But I no more my long-lost home shall hail!" (ll. 2, 9, 12, 14). In this way, Williams's preface and translation reshape Bernardin's text, asserting originality through sonnets of sensibility and natural history that reinvigorate her poetic reputation and profess ties to Britain as well as victimization under Robespierre's government. She alters the novel with personal and political agendas in view, appealing to Britons critical of her continued support of revolutionary ideals, habitation on the continent, and political involvement. After her release from captivity in November 1793, Williams claimed that her personal danger escalated as excerpts of her accounts from France appeared in British newspapers read by Robespierre's committee of public safety in Paris or, as Williams called it, "the committee of public extermination" (2:2:5). Although Patrick Vincent and Florence Widmer-Schnyder argue that she and her companions secretly may have been on a mission for this committee, Williams depicted her life in peril and traveled to Switzerland, a destination that afforded materials for the translation in which she most prominently incorporates her knowledge of natural history.[36]

RAMOND AND DARWIN: GENDER POLITICS AND TRANSLATING THE GLACIER GODDESS

Williams journeyed to Switzerland in 1794 but did not publish her *Tour in Switzerland* until 1798, after Robespierre's overthrow and her return to Paris with renewed hope that Napoleon's empowerment would revive principles of liberty in France.[37] With steadfast commitment to French revolutionary thought, she critiques "the present state of the governments and manners" of Swiss cantons in the context of travel writing about the Alps. In the main text of her *Tour,* Williams only casually references natural history, mentioning frequent stops "to botanize" the "rich variety of herbs and delicate mountain-flowers," and just as she disclaimed political knowledge in *Letters,* she claims ignorance of mineralogy while discussing "quartz, mica, and schorl." Circumspectly placing herself within the proper bounds of women's scientific knowledge, she describes mineralogical discourse as "not being perfectly intelligible to me," as opposed to the "naturalists" of her party, who "marched off to examine whether an adjoining mountain had most strata of white feltspar or green granite" (*T,* 1:184–85). Her conscious maintenance of feminine propriety while engaging with mineralogy becomes explicit when visiting an abbot who shows her his collections of artificial flowers and of minerals. Williams explains, "as a female the Abbot ought to have given me a nosegay of flowers, but, thinking probably the present more portable, he presented me with two very fine specimens of the purest rock-chrystal" (*T,* 2:105). She here establishes knowledge of flowers as more acceptably feminine than that of mineralogy and geology, which David R. Oldroyd describes as "a quintessentially manly or sportsman-like kind of science."[38]

Although she hesitates to trespass too noticeably against feminine decorum in the text of her *Tour,* Williams demonstrates bolder scientific participation in her work's appendix, where she translates a technical study of the Alps, Alpine glaciers, and contemporary geological theories. Interpreting Louis-François Ramond, baron de Carbonnière's "Observations on the Glacieres, and the Glaciers," she creates original footnotes to supplement his scientific treatise.[39] In addition to his labors as a botanist and geologist, Ramond had become an elected deputy of Paris in 1791, where he tried to calm Jacobin enthusiasm and was forced to flee to the Pyrenees in August 1792. Williams and Ramond likely met in Paris sometime between 1790 and 1792, and corresponded thereafter.

Augmenting her translation of Ramond's work with her own observations, Williams extends his notes so that, at times, it becomes difficult, if not impossible, to distinguish her scientific interjections from his without comparison to Ramond's original edition. The translation's three longest footnotes belong, in part or in entirety, to Williams. The first of these comprises an excerpt of a letter from Ramond, updating his research (*T,* 2:284–85). Establishing herself as Ramond's scientific correspondent, Williams illustrates that she possesses his approval of her translation as well as sufficient knowledge to discuss geological subjects with geologists themselves.

In his essay, Ramond exemplifies the Alps to dramatize a geological dispute about the earth's formation, pitting the so-called Neptunists against the Plutonists or Vulcanists. He associates Neptunism's water-based theory with Horace Bénédict de Saussure, and the heat-oriented Vulcanist hypothesis with Buffon (*T,* 2:287). Significantly, Williams joins this debate in her second interpolated footnote to Ramond's text and, while Ramond draws on Buffon to represent Vulcanism, Williams, writing eleven years later, is indirectly influenced by a more current proponent of the theory, James Hutton. Advocating existence of the earth's central heat and its agency in the globe's past and future changes, Hutton has been called "the founder of modern geology";[40] he published his *Theory of the Earth* in 1788 and a much expanded version in 1795. Whether or not Williams read Hutton's texts, she certainly received his ideas from another scientific source, Erasmus Darwin, Hutton's friend and correspondent, who cites Hutton's ideas in his *Economy of Vegetation* (1791), the first part of his long scientific poem *The Botanic Garden.*[41]

In her footnote, Williams critiques Darwin's espousal of Huttonian thought in explaining higher altitudes' colder climates. Ramond remarks, "Some who attribute to the earth an *absolute* heat . . . suppose that the mountains, from being insulated masses distant from the central focus, are subject to a greater loss of internal fire" (*T,* 2:291). While Ramond identifies no adherents to this theory, Williams cites Darwin's *Economy of Vegetation,* a text with which women writers less frequently engaged than the more botanically focused second part of *The Botanic Garden, The Loves of the Plants.* Perhaps Darwin's quotation of Williams's *Letters from France* in a note to his second Canto of *Economy of Vegetation* helped prompt her attention to this broader study of geological and meteorological phenomena that affect botanical species. Quoting from Darwin's notes on springs and on glaciers, Williams relates that, for Darwin, "the primary cause, why the summits of mountains are much colder than the plains is, their being in a manner insulated, or cut off from the common heat of the earth, which is always forty-eight degrees, and perpetually counteracts the effects of external cold beneath that degree" and that "the snow which lies in contact with [the earth] is always in a thawing state . . . hence, in Italy, considerable rivers have their source from beneath the eternal glaciers, or mountains of snow and ice" (*T,* 2:291–92).

According to Williams, Darwin's explanation contains inconsistencies, and she challenges him to elucidate this Huttonian view, questioning "how, since the summits of the higher Alps are insulated, and cut off from the action of the central fire, it could melt, by its heat, the eternal glaciers into considerable rivers" (*T,* 2:292). Skeptical of Darwin's narrative detailing geological processes responsible for glacial melting and climatic alterations in different altitudes, Williams shifts her critique to the more feminine medium of verse.[42] She interjects in the footnote her thirteen-stanza poem "The Complaint of the Goddess of the Glaciers. To Dr. Darwin."[43] Her Glacier Goddess represents a revision of Darwin's poetic deity, the Goddess of Botany. Visually inverting Darwin's technique of supporting his imaginative poetry with extensive scientific prose footnotes, Williams interpolates her poem as a long footnote that crowds out her translation of Ramond's scientific prose.

In Williams's poem, the speaker, a native of Britain, recounts that the "Glacier-Goddess" approached her on "the Alpine cliff" and charged her with revealing to Darwin the Alps' natural mysteries (ll. 13–14, 4). Acknowledging Darwin's Rosicrucian investigation into the four elements of earth, air, fire, and water, the goddess advertises "new marvels" and "treasures" in her frigid "realms" (ll. 18, 20). These "treasures" manifest economically in, for instance, mineral stores, as well as intellectually in the goddess's offer to correct Darwin's inadequate note on the origin of Alpine "springs" and to "unlock the rivers viewless source" (ll. 43–44, 39–40). Williams thus playfully proposes (through the Glacier Goddess) to educate Darwin in scientific matters.

Championing the frigid climate and complaining of misrepresentation, Williams's goddess additionally takes Darwin to task for lines from *Economy of Vegetation* that demonize agents of the cold, asking, "Ah, why a vestal to a [male] 'fiend' transform?" (l. 25). Darwin's poem envisions a natural battle in which his Goddess of Botany "calls her hosts to arms," ordering her nymphs to "unite" and to

> Call your bright myriads, trooping from afar,
> With beamy helms, and glittering shafts of war;
> In phalanx firm the FIEND OF FROST assail,
> Break his white towers, and pierce his crystal mail. (Canto 1, ll.
>   437–40)

In the context of his botanical poem, Darwin unsurprisingly depicts frost as a "fiend." He notes, "the principal injury done to vegetation by frost is from the expansion of the water contained in the vessels of plants . . . which are distended and burst," ending in plants' destruction.[44] Williams converts Darwin's masculine "FIEND OF FROST" into the feminine Glacier Goddess, who assumes the role of wounded virtue. Both proud and seductively vulnerable, her goddess assures Darwin that the hand beckoning him also "impels/The rushing Avalanche" and can "transfix thee numb'd, in icy cells" (ll. 29–30, 31).

In the end, the Glacier Goddess's displays of feminine knowledge and power fail to attract Darwin, and her illuminations of "new marvels" stall out midline, as she suddenly realizes she's lost, or rather never had, his attention, lamenting, "For thee—but ah, my pensive form he flies/For nymphs of golden locks, and florid hue!" (ll. 49–50). Williams's speaker concludes the poem with a sympathetic last look at the goddess, who then "wept, and folded in a cloud, withdrew" (l. 52). The poem's comical elements lighten Williams's depiction of female expertise rejected by a male figure of accepted scientific authority. Her portrayal reminds that Darwin's own poem imaginatively objectifies and conflates women and nature (especially flowers) so that both become subject to his study and interpretation. She underscores the hypocrisy in a gendered structure of science where, when the tables are turned and women (and nature) attempt to instruct, they are ignored. Extending her role as Ramond's translator to that of original contributor, Williams chides Darwin for not attending more carefully to these cold regions even as his neglect facilitates her opportunity to explore these subjects in the poet-naturalist mode, making her own scientific assertions and queries, and inviting Darwin to reconsider or at least explain the seeming contradictions in his theory of what causes glaciers to melt sufficiently to form "considerable rivers" (*T,* 2:292).

Williams's depiction of Darwin refusing instruction from the Glacier Goddess anticipates critical dismissal of her own participation in science. Without citing evidence of errors in her translation of "Observations," a critic from the *Monthly Review* deems her incompetent in the realm of science and relegates her to that of literature, remarking, "She makes amends for these defects . . . by an address from the Glacier Goddess to Dr. Darwin."[45] According to the critic, Williams's "vigour of fancy" disqualifies her from both science and politics, and "Politics seem to be Miss W.'s favourite science, but it is not the subject in which she is the best qualified to excel."[46] In her *Tour,* Williams suggests that Swiss cantons would benefit from a revolution under the auspices of France. The British critic responds with an unlikely lumping of Williams with Burke, and states that "poetical politicians" are "objectionable" "since all sound moral and practical reasoning, to which the science of politics eminently belongs, is totally incompatible with the giddy flights of an unrestrained and impassioned fancy."[47] Insisting on the "incompatib[il-ity]" of literature with science as well as politics and refusing to follow the radical thoughts of this "female reformer," the critic finds her "too often Gallic" and disapproves of her enthusiasm for democracy.[48] However, while most reviewers felt at odds with Williams's politics, some, like the critic from *European Magazine,* admit that Williams's translation of "Observations" contains "many acute and philosophical reflections on the phenomena of nature."[49] Significantly, her contemporaries also would have recognized that her revolutionary politics resonate with geological discourse.

RAMOND AND DARWIN: GEOLOGICAL REVOLUTIONS AND COSMOPOLITAN CLIMATOLOGY

The dispute between Neptunism and Vulcanism anticipates, in Williams's translation of Ramond, the contention later known as that between "catastrophism" and "uniformitarianism" to explain how changes have occurred on the globe since its formation.[50] While, for Ramond, the key players in this debate remain Saussure and Buffon, Williams again updates the controversy with respect to Hutton's considerable influence, as absorbed through Darwin. Hutton formulated a cyclical theory of the earth, in which lavas are "successively poured forth and then eroded away, giving a landscape that was continually changing in detail through time, but remaining much the same overall."[51] Allowing for occasional cataclysmic "accidents," Hutton's ideas of gradual change and general uniformity came to be known as uniformitarianism, and countered the catastrophist notion that the earth's changes owed primarily to periodic violent events followed by little subsequent change. Interestingly, in discussing alterations to the earth, both sides of this dispute employed the term "revolution" as a signifier of crustal change affecting a whole region, especially in the formation of mountains.[52] Diderot and d'Alembert's *Encyclopédie* describes geological revolutions as "the natural events by which the face of our globe has been and is still being continually altered in its different parts by [the action of] fire, air, and water."[53]

Of course, the term "revolution" had significant political, as well as geological, implications in the second half of the eighteenth century. According to Alan Bewell, "By the 1790s, geology had assumed the status of *the* preeminent science of revolution," a claim earlier made by Ramond.[54] Buffon directly relates revolutions in political history to those within natural history, beginning his *Epochs of Nature* (1779), stating, "Just as in Civil History, one consults titles, searches for medals, and deciphers ancient inscriptions in order to determine the epochs of human revolutions and to establish the dates of human or civil events, so in the same way in Natural History, it is necessary to dig into the archives of the world, drawing ancient monuments from the entrails of the earth, collecting their debris, and gathering together in one body of proofs all the clews [*sic*] of physical changes which can enable us to regain the different ages of Nature."[55] Cuvier would later echo this connection between sociology and geology; and, indeed, the debate over the concept of revolution sometimes formed along political lines, so that British uniformitarians like Hutton and Lyell could look to the Glorious or Bloodless Revolution as a very different model for change from the terror-filled revolution witnessed by the catastrophist Georges Cuvier in France. At stake in this dispute between directional or noncyclical catastrophism, and the cyclical gradualism of uniformitarianism was the length of time and degree of violence necessary for the earth's alterations to occur.

In the Swiss Alps, Ramond views the progress of glaciers as a case study in gradual change and the mountains as testimony to the great age of the earth in "the epochas of its revolutions" (*T,* 2:282–83).[56] He observes that avalanches' and glaciers' downward movement exposes excessive snows and ice "to the action of the heat" that soon melts and "re-establishes the balance between the loss and the increase" of snow (*T,* 2:336–37).[57] Taking up this notion of balance and of the gradual, rather than violent, progress of glaciers Williams expands one of Ramond's footnotes to encompass her ideal of peaceful political revolution and the notion that humanity can bring improvement by shaping the changes or revolutions in nature.

In her earlier *Letters from France,* Williams correlated political change with enlightenment improvement to nature, as endorsed by naturalists including Buffon. For him, success in improving nature denotes a nation's degree of civilization, and such efforts to cultivate nature could even amend that nation's climate.[58] Exemplifying British political and environmental reformations as models for France, Williams declares, "May Liberty, which for so many ages past has taken pleasure in softening the evils of the bleak and rugged climates of the North, in fertilizing a barren soil, in clearing the swamp, in lifting mounds against the inundation of the tempest, diffuse her blessings also on the genial land of France" (1:1:25). However, by the time of the Reign of Terror, Williams more darkly associated the French Revolution with geological and meteorological phenomena, so that now "the political clouds . . . gathered thick around the hemisphere: we heard rumours of severity and terror, which seemed like those hollow noises that roll in the dark gulph of the volcano, and portend its dangerous eruptions" (2:1:5).[59] Her portrayal shifts from the human potential to improve nature (and thus society) to an image of catastrophic revolutions in nature, emanating from destructive political forces. Only after the fall of Robespierre does Williams, in her 1798 *Tour,* renew her hopeful vision of revolutionary ideals and positive human control over political and natural realms.

Appropriating Ramond's footnote on Alpine glaciers, Williams radically asserts the human capacity to change nature and achieve global climatic balance. Once more referencing *The Economy of Vegetation,* she enthuses, "Darwin, alarmed by the increase of the ice at the northern pole, and on Swiss mountains, has pointed out an ingenious and beneficent mode of restoring the equilibrium of heat and cold by employing the fleets of Europe, now busied in devastation, in the more innocent occupation of navigating ice islands [icebergs] from the neighbourhood of the pole, to cool the feverish climates of the track of the sun" (*T,* 2:314). Darwin's "alarm" about the globe's eventual "refrigeration" finds its source in Buffon's well-known prediction. Conceiving the initial formation of the earth as a molten state, Buffon hypothesized that the planet has been in a gradual process of cooling that will persist until it becomes uninhabitable and dead.[60] This prediction continued to haunt the second generation of Romantic poets, most famously in Byron's "Darkness." And Percy Shelley, gazing on the Alps, "these palaces of death and frost," in the summer of 1816, wrote, in an often-quoted letter to Thomas Love Peacock, that he rejects Buffon's idea of universal cooling, but easily envisions local devastation through the glaciers' continual "increase of ice."[61] His sentiments are thus compatible with those of Williams, who eighteen years earlier witnesses evidence of glacial encroachments in the mountains' surrounding vales but refuses to view this threat on a global scale and is convinced instead by Ramond's "endeavours to prove that the Glaciers of the Alps, like those of the Pyrenees, are not susceptible of any durable increase beneath the icy zone" (*T,* 2:315).[62]

Although Williams repudiates Buffon's and Darwin's scientific anxieties about global cooling, she lauds Darwin's suggestion that "the fleets of Europe" should bring universal improvement rather than violence and death to humanity. Elaborating Darwin's idea of using icebergs to equalize the global climate, Williams notes that if Ramond had not disproved "the threatened progressive refrigeration of the globe from the increase of the ice on the Glaciers," then, since "the continental armies are about to cease their work of death, [Darwin] might also have proposed their being engaged in the removal of the Glaciers of St. Gothard, or Mont Blanc" (*T,* 2:314). She interpolates into her translation revolutionary fervor with this wish that "continental armies" would cease to war and pursue instead the democratic and humanitarian ideals of achieving a temperate, universal climate. It is difficult to overstate the significance of this suggestion since, in Enlightenment determinism, propounded by various naturalists and perhaps most famously by Montesquieu, a nation's climate dictates not only body type and skin color but virtually every aspect of inhabitants' identity, from temperament to sexual potency.[63] In this formulation, to universalize the climate would eradicate national differences, potentially producing a globally homogenous nature and nation, an unvarying human race of worldwide equality.

Yet, if this hope for a uniform, global climate cannot be realized, Williams, through Ramond, also relates a different kind of universalism; for the Alps, it seems, contain all the climates of the globe. Whereas the leveling cosmopolitanism that Williams associated with Cook arguably aligns with Darwin's call for a universal climate and absolute (in theory, at least) equality, Ramond additionally presents another kind of cosmopolitanism already existing in nature. Within the Alps, Ramond perceives a microcosm of the world due to the increasing altitude, explaining that "in a walk of a few hours you have felt the influence of all the seasons; seen the production of every climate; ran through the whole scale of vegetation, and compared the birds of Italy with those of northern lakes and continents," including North America (*T,* 2:345). He views these various species as analogues for humanity, finding "a family of birds, which is the emblem of our own; a republic of insects, which recalls our idea of our nations; their industry, their relations and antipathies" (*T,* 2:334). In such correlations between natural and social behaviors and in describing this space as containing "the production of every climate," Ramond's works influenced the rising naturalist Alexander von Humboldt, who later befriended and collaborated with Williams, and whose travels to South America would emblazon in the public imagination images of a mountain's global nature.[64]

Translating Humboldt: Chimborazo and La Physique Générale

In 1807 Humboldt published in French his *Essay on the Geography of Plants,* the first effort to capture in print his 1799–1804 expedition to the Americas in which he was accompanied by the French botanist Aimé Bonpland and received support from the Spanish monarchy. The account of this voyage, so rich in imaginative experience and scientific theories, collections, and measurements, would consume Humboldt's private fortune and countless hours in composition until his death in 1859. Williams translated into English two of Humboldt's chief works, the shorter *Researches, Concerning the Institutions and Monuments of the Ancient Inhabitants of America* (1814) and the seven-volume *Personal Narrative of Travels to the Equinoctial Regions of America* (1814–29), both of which expound on these early explorations, but the *Essay* remained untranslated into English until this past decade. Like these later texts, the essay created a sensation in both scientific and literary circles, and it became particularly recognized for its *Geography of Equatorial Plants: Physical Tableau of the Andes and the Neighboring Countries,* the work's centerpiece. This color profile of Chimborazo cuts away the entire right half of what was then thought to be the world's tallest mountain to present a tabula rasa upon which to record the Linnaean, and some indigenous, names of plants growing at different altitudes. In his depiction of Chimborazo, Humboldt "provoked people to think about the globe in fundamentally new ways—as a single entity with interlinked biological, physical, and cultural properties varying latitudinally and altitudinally in a systematic and comprehensible fashion."[65]

Humboldt sustains this attempt to encompass an understanding of universal nature through his *Personal Narrative,* laying the groundwork for what would later be termed "Humboldtian science" and would characterize a major current in nineteenth-century natural inquiry. In conceptualizing his approach to nature, more than any other source, Cook's scientific voyages most directly inspired Humboldt's ambitions as an explorer, yet Humboldt also sought to differentiate himself from this admired predecessor. As Williams explains in her preface to the *Per-*

Alexander von Humboldt's *Tableau Physique* (1807), charting his various observations and measurements taken during his ascent of Chimborazo. (Kew Gardens)

sonal Narrative, by penetrating the interior of South America, Humboldt perceived new opportunities for fieldwork and scientific achievement that could not be accomplished through "sea-expeditions' like those of Cook.[66]

In his explorations, Humboldt wished to depart from the mode of collecting and identifying plants associated with Cook's first voyage and Joseph Banks. Although he considered the steps of collecting and classifying to be necessary, he often contrasted "descriptive" natural history with his own emphasis on geographical variations of plants and of other physical parameters.[67] Humboldt strove "to create what he termed *la physique générale*—the universal, synthetic science that would comprehend both the unity and the diversity of nature."[68] Part of accomplishing his universalist form of scientific investigation involved measuring everything possible, using the latest technical innovations in scientific instruments, a litany of which he carried with him through South America. Humboldt hoped this wide array of measurements could then be compared with similar measurements performed in other geographical locations around the globe, making real linkages between apparently disparate phenomena more readily perceptible. One recent scholar represents Humboldt's terrestrial physics as "natural history conducted in an observatory," with numerous observatories located around the world.[69]

In her preface to the *Personal Narrative,* Williams praises Humboldt's imaginative mode of science. His illumination of harmonies and aesthetics through natural inquiry aligns him with the German Romantic movement, and Williams affirms that nature "speaks in a voice . . . well understood by the mysterious sympathy of the feeling heart."[70] In December 1821 Robert Southey admiringly wrote, "Humboldt is among travelers what Wordsworth is among poets. The extent of his knowledge and the perfect command which he has of it are truly surprising; and with this he unites a painter's eye and a poet's feelings."[71] However, Southey's surprise at Humboldt's poetic style may owe much to the translation. Just as she interjected her own insights when translating Ramond, Williams also takes liberties in her translation of Humboldt, augmenting his "passionate enthusiasm." A modern editor of the *Personal Narrative* states that Humboldt's "French is curiously flat, scientific and modern" and "I was struck by the disparity between" Humboldt's writing and Williams's English translation.[72] According to him, Williams's "collaborati[on] with Humboldt" proves "faithful . . . except when Humboldt enthused—then his translator interpreted and exaggerated."[73] Williams publicizes Humboldt's approval of her interpretation, stating in her preface, "I have been encouraged by the care with which [Humboldt] has read most of my pages, and corrected many of my errors."[74] Her version of the *Personal Narrative* inspired Charles Darwin's own quest for scientific discovery and accompanied him on the H.M.S. *Beagle.* Reading Williams's translation, Darwin called Humboldt's style a "rare union of poetry with science."[75] When Williams embellishes Humboldt's prose, melding her own originality with that of this naturalist, she often responds to the "poetry" of his methods and ideas, striving to create a unitary vision of the world by synthesizing its various phenomena.[76]

Recent historians of science argue that "only in [Humboldt's] pictorial representation" of Chimborazo in the *Essay on the Geography of Plants* "did he truly succeed in consolidating an abundance of particulars into a few generalities."[77] In the essay's preface, Humboldt ambitiously demarcates his universalist claim, writing, "Here I bring together all the physical phenomena that one can observe both on the surface of the earth and in the surrounding atmosphere."[78] Describing his ascent of Chimborazo, he delineates how the forms of vegetation, mammals, birds, and insects change at different elevations. Humboldt compares the Andes mountain summits of Chimborazo and Cotopaxi, "one of the most active volcanoes in the Quito province," with those of Europe, depicting on the tableau that Cotopaxi, for instance, "is almost five times the height of Mount Vesuvius" and that Mont Blanc, the highest summit in Europe, reaches only to the lower limit of permanent snow in the Andes.[79] He describes European countries' plants as "socially organized," exclusionary, and xenophobic, while the vegetation of the tropics is various, "less uniform" and interestingly appears to harbor a social ideal of democracy, for "no one plant dominates over the others."[80]

Yet, despite Humboldt's portrayal of nature's superior sublimity in the Americas, and his commitment to democratic, egalitarian possibilities for society, he also echoes enlightenment ideas about racial determinism, denigrating the inhabitants of these tropical zones in comparison with northern (especially European) countries, famously stating, "The civilization of peoples is almost always in inverse relation to the fertility of the soil they occupy. The more difficulties nature presents, the more quickly mental faculties are developed," for "the civilization of our species makes more rapid progress in the northern regions than amidst the fertility of the tropics."[81] As Ángela Pérez-Mejía notes, "Humboldt was a conqueror of knowledge, not of lands," and the knowledge presented in his new geography made possible "the alliance between Europe and the creole class," associating barbarism "with the native American ethnic groups, and civilization with the Caucasian races from whom the creoles were directly descended."[82] In the context of revolution's terrifying aftermath in France, Humboldt's work displaces concepts of barbarism and degeneration within Latin America and maintains European intellectual and industrial superiority.

In South America, where little foreign scientific exploration had been pursued in more than sixty years, Humboldt found the perfect setting to test his approach to nature and realized that in modeling a new mode of scientific inquiry he modeled a new kind of naturalist. Highlighting his individualism in the *Personal Narrative,* he contrasts the difficulty of his own voyage with the relative ease involved: "When a government undertakes one of those maritime expeditions, which contributes to the knowledge of the globe," with "no obstacle to the accomplishment of it's [*sic*] purpose."[83] Humboldt alludes to Cook's archetypal, nationally sponsored explorations and dismisses the neutrality granted to Cook's ships during the American Revolution that Williams earlier emblematized in terms of science's equalizing and universalizing power.[84] Embracing his stance as a different kind of voyager, Humboldt depicts "passports" and protection from "belligerent powers" as unknown luxuries, for "far different is the situation of a private individual, who undertakes a journey at his own expense into the interior of a continent, over which Europe has extended it's [*sic*] system of colonization."[85] He thus seeks to cultivate interest in himself as a new, independent scientific persona, affirming, "the curiosity of the public" is "oftener fixed on the persons of travelers than on their works."[86]

In Williams's translation, Humboldt balances this self-image of a solitary scientific explorer with participation in eighteenth-century rhetorical traditions of scientific cooperation and sociability by referencing contemporary naturalists' works and his personal exchanges with them.[87] As Alison Martin has shown, Williams accentuates this collaborative mode in Humboldt's texts. For instance, when Humboldt places himself within the scientific community by describing his "constant friendship" with Georges Cuvier as "si honorable" and "utile," Williams translates this as "highly honourable" and "advantageous," promoting Cuvier's contributions to Humboldt's success.[88] In this way, she emphasizes Humboldt's accord with her earlier romanticization of Cook's collective, humanitarian values. In her preface, Williams hopes Humboldtian science may succeed where the French Revolution failed, declaring, "How often will posterity also turn from the terrible page of our history, to repose on the charm of a narrative, which displays the most enlarged views of science and philanthropy!"[89] Nevertheless, Humboldt's totalizing system based on differentiation and his self-aggrandizing portrayal of isolated genius create dissonance with Williams's ideal, as she subtly acknowledges in her poem on the younger explorer.

While Williams and Humboldt shared political liberalism, support of the American and French revolutions, and lifelong opposition to slavery, Humboldt additionally inspired a rhetoric of conquest, albeit "peaceful" conquest. Hailed as "the Napoleon of science" for aspirations to universal domination of nature, Humboldt's sobriquet denotes a different kind of "connecting zone" than that envisioned in Williams's poetic tribute to Cook.[90] Her short, personal poem "To the Baron de Humboldt, on his Bringing Me some Flowers in March" (pub. 1823) invites comparison with the humanistic science she celebrated in "The Morai":

Sooth'd I receive the flowers you bring,
Whose charm anticipates the Spring;
Whose tints in vernal freshness vie
With plants beneath an austral sky,—
Those glowing plants that, long unknown,
Your travell'd science made our own:—
Bright gift! in lavish grace array'd,
Thy flowers have only bloom'd to fade,—
Their transient being soon forgot:
How far unlike the giver's lot!

Williams's verses stage an aesthetic rivalry between the "vernal" European flowers given to Williams by Humboldt and the "glowing," "long unknown" flowers he "discovered" in South America. Her brief depiction of Humboldt's science as a process of discovering, collecting, and making "known" botanical species repeats the naturalism associated with Joseph Banks and the Cook voyages, exactly the kind of botanical study from which Humboldt sought to dissociate himself. Yet if, in this regard, Williams aligns Humboldt with Cook, she distances him from the humanism she attributed to the earlier explorer when she describes South American plants as possessions "your travell'd science made our own." Since Williams distinguishes between European and South American flowers, the "our" of this poem clearly refers to Europe, so that Humboldtian science contributes to the conquests making South American plants—and colonies—Europe's own. Signifying Europe, the poem's "our" paradoxically represents a form of exclusion rather than the universal inclusion Williams depicted in the Cook poem where the islanders, too, were included in the "connecting zone." There is no equal exchange here, only uni-directional appropriation. When Williams assures Humboldt of lasting fame, it is not a fame she envisions sung in the oral histories of people indigenous to South America, as had been the case with Pacific islanders in her verse memorial to Cook.

Humboldt's assumption of what Pratt describes as "a godlike, omniscient stance over both the planet and his reader" registers a changed mode of science from the natural history Williams associated with democratic values.[91] Provoking identification with both scientific col-laboration and solitary conquest, Humboldt implements methods of specialized research and science's professionalization, eliciting a cosmo-politanism that acquires generalizations through maintenance, rather than erasure, of particulars and differences, and this opposes the cos-mopolitanism achieved through the unity in equality and homogeni-zation Williams attributed to Cook and Darwin. The fact that Williams recognizes Humboldt's departures in a poem published separately, and does not insert the verses directly into her translation of his text as she had done in earlier collaborations with naturalists, also marks gradual changes in women's and literature's relationship to science.

Although she intervenes less noticeably in her translations of Hum-boldt's works than in those of Bernardin and Ramond, Williams's later appropriations arguably form her most important interventions be-cause they occur within the expected seriousness of Humboldt's scien-tific theories, explorations, and discoveries, targeting a wide-ranging audience of professional as well as amateur naturalists. Her interjections of original poetry and footnoted politico-scientific critique, so appar-ent in her earlier translations of naturalists, rework genres in which she could more comfortably distinguish her authorial presence: a sentimen-tal novel and an appendix to her own travel writing. Humboldt's tech-nical scientific details and experiences place his person and perspective at the center of his text, precluding prominent interpolations or com-mentary from his female translator. Nevertheless, Williams's preface and alterations to Humboldt's scientific research, especially in heightening his rhetoric of cooperation and poeticizing his language more generally, reveal her influence in this collaboration, even as this reduced visibility also elucidates reduced acceptance of public participation for women and literature within science's increasing specificity, elitism, and gravity of purpose. Whereas previous writers such as Seward and Smith confi-dently corrected naturalists' observations, as Williams does in her ear-lier treatments of Darwin and Ramond, science's and literature's grow-ing, separate professionalizations, as well as expectations of fidelity in scientific translation here require more subtlety in Williams's original interventions of creative and political thought. In this way, as science emphasizes greater technical specificity and individualism, Williams's authorial persona as translator conversely becomes less recognizable. For her, such shifts in science and literature again correlate with France's political climate.

BEYOND TRANSLATION: TOWARD THE PROFESSIONALIZATION OF SCIENCE

Williams's introduction to her *Poems on Various Subjects* (1823) censures assertions in the *Edinburgh Review* regarding France's "present degenerate State of Science and Literature." She remarks, "The professors of science in this country [France] may indeed be safely left to defend themselves. The learned only are fit to be their own judges, and I know not what my eulogium could add to such names as those of La Place, Delambre, Hauy, Cuvier, Jussieu, Gay-Lussac, Arrago, Biot, Thenard, and many others worthy to augment the list."[92] Yet, her praise also acknowledges the difficulties French science faced in recent, revolutionary years: "What, for instance, can be more noble and affecting than the conduct of Condorcet and Rabaut St. Etienne, at that period? who, while *hors la loi,* and certain, if their retreat were discovered, of being dragged without trial to the scaffold, pursued with the calmness of a superior nature the lofty speculations of philosophy, and left posthumous works, in which they disdained to make the slightest allusion to their own desperate situation, which for both terminated in death!"[93] Whereas Cook's example had allowed Williams to view various nations, and France in particular, as designating scientific thought as an international safe-harbor in political storms, the fallout of the Terror transformed science's relation to politics. Without (inter)national protection and, indeed, facing political persecution, these naturalists' "lofty speculations" in the midst of "their own desperate situation" and withdrawal from turbulent society helps to mark their "superior nature."

Williams's veneration of these naturalists' work in isolation complies with the developing cult of genius in the early decades of the nineteenth century in both literature and science. Although now free of political persecution, some members of the younger generation of naturalists Williams commends as belonging to "the new order of things" embraced this mode of individualism. Cuvier, for instance, strongly advocated specialized research and "viewed public interest as a possible danger to the advancement of science," fearing that efforts to capture the public imagination made naturalists imprecise in their methods and descriptions.[94] In this vein, Humboldt arguably mediates between a collective, collaborative mode of natural history and professionalized science. He "gave lectures, organized meetings, wrote letters by the hundred, visited dignitaries, held forth tirelessly . . . in salons" like that of Williams, and believed that a history of the earth, a geology, could only be achieved by equipping the general public with precise and easy-to-read instruments, allowing them to participate in realizing his universalist form of scientific inquiry.[95] His all-encompassing, comprehensive enterprise at once countered and helped inspire the division of science into more specialized disciplines. Although Humboldt's personality and approach to nature highly influenced scientists throughout the nineteenth century, his popular appeal, especially to women readers such as Williams, did not accrue everyone's respect.

When, in 1804, upon their only meeting, Napoleon famously and derisively told Humboldt that his wife, too, was fascinated with botany, the French emperor intentionally feminized both natural history and the naturalist by emphasizing women's interest in the field.[96] Byron similarly discussed Humboldt's discoveries' effects on scientific-minded women, writing in the fourth Canto of *Don Juan*,

> Humboldt, 'the first of travelers,' but not
> The last, if late accounts be accurate,
> Invented, by some name I have forgot,
> As well as the sublime discovery's date,
> An airy instrument, with which he sought
> To ascertain the atmospheric state,
> By measuring "the *intensity of blue:*"
> Oh, Lady Daphne! let me measure you! (4:889–96)

Byron refers to one of the numerous instruments Humboldt carried through South America, the cyanometer, "invented" by H. B. Saussure, not Humboldt; it measures the color intensity of blue sky to indicate transparency and water-vapor levels. Byron thus satirizes the instrument's capacity to measure the "*intensity*" of a bluestocking. For him, Humboldt's mania for measuring, and natural history's association with women, made this science laughable. In this climate of ridicule, increasing efforts to bar women from specialized knowledge through its retreat into universities and scientific societies, very different from Williams's inclusive salons, aided in apotheosizing the isolated male scientist.[97]

Certainly, Williams never endorsed the masculinization of knowledge, yet, over the course of her career, she reflects this politico-scientific shift in perceptions of universal humanism and individualism. Republishing "The Morai" in her 1823 *Poems,* she adds a note explaining the verses' original inclusion in Kippis's *Life of Cook* and recounting alterations in personal circumstances that have occurred since that time. Grieving for this father figure, Kippis, who died in 1795, Williams now affirms, "Nothing, indeed, is better fitted to confirm our love and admiration of particular virtue, than experience of the world in general."[98] This constitutes a sobering and startling statement at the end of a poem celebrating "the world in general," that is, humanity "in general." For Williams, in a way reminiscent of Humboldtian principles, "experience" teaches her to value particulars, differences, individuals at least as much as notions of universality, exemplifying parallel trends in both science and literature toward a projected sense of individualism.

Nevertheless, another contemporary woman writer warned against the dangers posed by the individual, solitary male scientist. While Byron mocked the feminization of natural history, Mary Shelley's *Frankenstein* (1818) more seriously satirized science's masculinization and the horrifying potential of discoveries (and even procreation) in isolation;[99] she thus reproves natural history's growing divestment both of women's influence and of its collective, collaborative mode. My next chapter examines how Shelley's later novel *The Last Man* (1826) participates in geological theories of species' origins as well as endings, creating literary originality by assessing naturalists' extinction hypotheses. In doing so, she critiques both Romantic individualism and scientific generalizations, and employs stylistic techniques that reveal further, developing tensions in women's scientific literature.

6

Reconstructing Origins

The Psychologization of Geological Catastrophe in Mary Shelley's *The Last Man*

In her third novel, *The Last Man* (1826), Mary Shelley intermixes fact and fiction, recasting the question faced by Helen Maria Williams of how to achieve originality when translating scientific texts and historical events. Inviting autobiographical identification between herself and the "author" of the novel's introduction, Shelley explains that her account results from translating Sibylline leaves inscribed with prophecies that detail humanity's future extinction by a devastating pandemic. Within the novel's frame, she thereby connects literary and biological "texts" as well as their origins and endings. Projecting human extinction into the future, Shelley wrote at a time when speculation about past extinctions became especially prevalent due to new discoveries in developing disciplines of geology, paleontology, and comparative anatomy. As geologists sought to reconstruct fossilized remains of extinct species, their efforts to understand how and why such extinctions occurred unearthed more questions than answers. Just as her introduction temporally looks both backward and forward, Shelley explores various contemporary theories about what may have caused these past species' demise as a means of guiding her narrative anticipating human destruction. She writes in the scientific mode of such earlier women writers as Charlotte Smith but is more subtle in her corrections and challenges to naturalists' assertions. Highlighting parallels between her interpretive work as authorial "translator" and geologists' interpretations of the earth and its fossil record, she critiques these scientific theories and their implications.

Shelley's geological interventions in *The Last Man* primarily center on the catastrophist theories of Georges Cuvier and William Buckland, but do so differently from previous works within the last-of-the-race genre. In her text's concluding pages, the main character and sole human survivor carves into stone the inscription, "2100, last year of the world!"[1] The novel, however, offers no indication that this marks the last year of the world, merely the last year of humanity. The earth, all of nature, continues on, unaffected by humankind's extinction. As other scholars note, Shelley published during this genre's waning popularity, following poems such as Jean-Baptiste Cousin de Grainville's *Le Dernier Homme* (1805), Lord Byron's "Darkness" (1816), and Thomas Campbell's "The Last Man" (1823) that depict humanity's annihilation as apocalyptically corresponding with the death of the sun and of every living thing on the earth.[2] Many critics trace these images of cataclysmic destruction to hypotheses of Georges-Louis Leclerc, comte de Buffon and Cuvier. At the end of the eighteenth century, a time when most naturalists rejected the notion that God would allow a species of His own creation to be eliminated, Cuvier established extinction as incontrovertible fact. He attributed its cause to past instances of sudden, catastrophic changes on the surface of the globe, and his work on fossils influenced geological thought throughout the early decades of the nineteenth century.

Since Shelley imagines the extinction only of the human species, rather than the catastrophic destruction of species on a grand scale, even critics conscious of Cuvier's influence on the last-of-the-race genre refuse to consider her novel within a Cuvierian context because, in Shelley's conception, humankind ends, so to speak, with the whimper of plague, not the bang of geological catastrophe.[3] In contrast, I argue that her portrayal of humanity's extinction remains consistent with Cuvier's directional notion of geohistory. Because, in his thinking, the magnitude of catastrophes produced by the earth gradually weakens over the course of time, Cuvier's theory cannot apply to humanity's future demise. Shelley represents this planetary enervation through anticlimatic natural disasters and by radically shifting geological catastrophism into the psychological "world" of the individual. In this way, Cuvier's hypothesis of how extinctions of past species occurred becomes one of the many that Shelley invokes only to dismiss in positing her own extinction theory through plague.

Despite Shelley's focus on extinction and her acclaim as the Romantic era's woman writer most famously associated with science through her first novel, *Frankenstein* (1818), recent scholars overlook geological references in *The Last Man,* instead concentrating on issues of empire, politics, gender, and disease.[4] Interestingly, the novel establishes these themes in ways that correspond with contemporary geological debates. Shelley's microcosmic shift, her psychologization of catastrophe, portrays individuals as "worlds," as she examines how individual deaths affect larger public and domestic relationships, creating a geological framework in which the extinction of even a single life constitutes apocalypse. Her attention to the world of the mind privileges authorship as well as a middle ground between two polar extremes: abstraction and excessive individualism. While Shelley critiques scientific generalizations that ignore the fate of individuals, such as the homogenizing notion of species extinction, she also interrogates lastness by associating this concept with Romantic interiorization that takes individualism to the point of isolation. Classic readings of Wordsworthian interiorization by scholars such as M. H. Abrams, for instance, here help demonstrate how Shelley revises contemporary perceptions of individual experience. At the same time, her novel's less technical engagements with science exhibit a stylistic shift away from the more obviously intertextual techniques of earlier women's scientific literature and toward now-conventional ideas of Romantic solitude and autonomous originality. In *The Last Man,* these extremes, of abstraction and individualism, alter characters' physical and psychological worlds, dramatizing geological discourse.

Geology as Authorship: Re/Constructing the Past

For Mary Shelley, in the years immediately preceding publication of *The Last Man,* personal loss seemed a way of life. Between 1814 and 1819 she buried three children, having only one child survive to adulthood. Her husband, Percy Shelley, drowned in July 1822, and already writing her novel, the night before learning of Byron's death in 1824, she famously recorded in her journal, "The last man! Yes I may well describe that solitary being's feelings, feeling myself as the last relic of a beloved race, my companions, extinct before me."[5] Her presentiments of the "coming event" of Byron's death signify a kind of prophetic dread that pervades the emotions of characters in her novel. Shelley acknowledged her autobiographical intention of recreating in this book Percy Shelley as Adrian, the Earl of Windsor, and Byron as Lord Raymond.[6]

She casts herself as a combination of the mournful widow Perdita and Lionel Verney, who functions as narrator and epitomizes loss and isolation as the last man. Projecting this "beloved race" into an imagined future, Shelley begins the main plot of her novel shortly after England's nonviolent transition from monarchy to a new republic near the end of the twenty-first century.

Although other scholars note the political significance of setting this new republic three hundred years after the French Revolution, Shelley also reflects her era's preoccupation with geological revolution.[7] In the early nineteenth century, Cuvier, as the world's foremost authority in the new fields of paleontology and comparative anatomy, situated the extinction of past species within violent "revolutions" or changes of the earth. Shelley's self-identification as "the last relic" of an "extinct" race formulates her feelings of personal loss in relation to contemporary geological concerns. Significantly, although the Shelleys knew of Cuvier's work earlier, they ordered his masterpiece of fossil studies, *Recherches sur les ossemens fossiles* (1812), only a few weeks before Percy's death its arrival, then, would nearly accompany Mary Shelley's loss of her husband, a bereavement on which she dwells obsessively in her journal during the years of writing her roman à clef.[8]

In her introduction, Shelley oscillates between identifying the novel as factual and imaginative. She preserves the illusion of reality, in part, by explaining that her knowledge of future events results from an excursion, corroborated by her personal letters and journal, to "Naples in the year 1818" (*L*, 1). There, the introduction relates that she and her companion (implying Percy Shelley) "entered the gloomy cavern of the Cumæan Sibyl" and purportedly discover the ancient sibyl's prophecies: "piles of leaves, fragments of bark, and a white filmy substance," all "traced with written characters" in various languages, communicating the story of humanity's extinction (*L*, 2–3). Shelley alludes to Percy's death, revealing that her "selected and matchless companion" is now "lost to me." Finding consolation "in deciphering these sacred remains," the sibyl's leaves, she translates and arranges these "truths" (*L*, 3, 4). However, she also refers to her account as "fictitious," for "scattered and unconnected as they were, I have been obliged to add links, and model the work into a consistent form" (*L*, 3–4). Indeed, she declares, "Sometimes I have thought, that, obscure and chaotic as they are, they owe their present form to me, their decipherer" (*L*, 4). Translating her narrative from these organic materials and adding original elaborations, she admits they may be interpreted differently, emphasizing the subjectivity of her work.

Setting her cave exploration among the "antiquities" of Naples and the ruins of Pompeii and Herculaneum, Shelley invokes the common contemporary analogy between studies of past epochs of the earth and of past civilizations occurring in the opening of Cuvier's "Preliminary Discourse" to *Ossemens Fossiles*. He describes fossils as "monuments," declaring, "As a new species of antiquarian, I have had to learn to decipher and restore these monuments, and to recognize and reassemble in their original order the scattered and mutilated fragments of which they are composed."[9] Through this process of deciphering and reordering "scattered and mutilated fragments," the naturalist recovers the earth's historical narrative.

Cuvier's mode of deciphering nature largely departed from that of his scientific predecessors, and he presented species' extinctions and their fossils as key to discerning the earth's geological past. In his 1796 landmark paper on the subject, he employed comparative anatomy to demonstrate that African and Indian elephants constitute different species, and that the fossilized remains of mammoths belong to a third, separate species, while the "Ohio animal" (or mastodon) composed an entirely separate genus.[10] Cuvier determined that the greater their depth in the earth's strata, the greater the age of recovered fossils, and proposed that by examining and reconstructing "these monuments," the various revolutions or changes of the globe may be assessed. He suggested that the extinction of the mammoth and mastodon, as well as fossil species of, for instance, rhinoceros, bear, deer, and crocodile, with no living analogues, "prove the existence of some world previous to ours, destroyed by some kind of catastrophe."[11] Following Buffon in claiming that a succession of cataclysmic revolutions divided the history of the earth into six epochs or periods, equated with the "days" of creation in Genesis, Cuvier viewed these revolutions as geological processes by which dry land emerged from the sea to form new continents while old continents sank beneath sea level. An advocate of species fixity, he asserted that each revolution forced an affected set of the earth's fauna into extinction, only to be replaced by a "new" group through migration from a different part of the globe, determining that presently extant species represent populations of humans and animals that survived from the previous continents, now beneath the sea.[12]

Cuvier's antiquarian rhetoric, likening the study of comparative anatomy to that of fragmented monuments, gestures toward the vision necessary to scientific thinkers recovering these origins and imaginatively reconstructing the past, an endeavor connecting not only with Shelley's setting among the ruins in Naples, but also, of course, with the sibylline leaves that she must, like Cuvier, imaginatively decipher and "reassemble in their original order" (*L*, 4). Presenting fossil reconstructions as crucial to understanding the earth's past worlds, Cuvier boasted that through his knowledge of comparative anatomy he could rebuild an entire fossil skeleton from a single bone, an important skill since often only a few bones or fragments survived from the original anatomical frame. Similarly piecing together "scattered and mutilated fragments," Shelley's reconstruction of sibylline prophecies represents an act analogous to that at the center of her first and more famous novel, which some paleontologists viewed as a horrifying correlative to their work.[13] While erecting the fossilized remains of an iguanodon, the geologist Gideon Mantell found himself, "like Frankenstein, . . . actually appalled at the being which rose beneath his meditations."[14] Shelley's material depiction of pestilence in *The Last Man*, which leaves a multitude of unburied human corpses scattered over the earth to decompose and perhaps become fossils themselves in time, led a critic from the *Monthly Review* to complain, "It is not a picture which she gives us, but *a lecture in anatomy*, in which every part of the human frame is laid bare to the eye."[15] The visionary power associated with Cuvier's comparative anatomy, his recovery of horrific beings and events from the earth's history, creates a context in which Shelley's reconstructed fragments from the past envision an equally horrifying future for humanity.

BYRON, CAVES, AND GEOHISTORICAL PROPHECY

Mary Shelley's tacit intertextuality, her use of Cuvier's work as a means to understanding past and future worlds, originated out of her readings not only of *Ossemens Fossiles*, but also of the poetry of Byron, and especially his closet drama *Cain: A Mystery* (1821). Interestingly influencing Shelley's novel, *Cain* appropriates geohistory's visionary capacity and challenges geologists' theological claims, illuminating scientific controversies that threatened to destabilize naturalists' (and women writers') traditional justifications of their scientific studies through natural theology. In publishing *Cain*, Byron acquired the enmity of many geologists for exposing Cuvier's ideas to attack from biblical literalists. Byron's

"Note" to the preface emphasizes geological support for scripture, explaining, "The author has partly adopted in this poem the notion of Cuvier, that the world had been destroyed several times before the creation of man. This speculation, derived from the different strata and the bones of enormous and unknown animals found in them, is not contrary to the Mosaic account, but rather confirms it."[16]

However, in the drama's ensuing dialogue, Byron's Lucifer employs geological findings as a goad, encouraging humans to question and turn from God, and thus associates geology with blasphemy. Conjuring up Cuvier's theory of successive revolutions of the earth, Byron's Lucifer refers to God alternately as "the Destroyer" and "the Maker," for "he makes but to destroy" (*C,* 1.1.265–67). Revealing to Cain "the history/Of past, and present, and of future worlds," Lucifer distinguishes Adam as only the first of the latest race of earth-dwellers, and explains that "mightier things have been extinct/To make way for much meaner than we can/Surmise" (*C,* 2.1.24–25; 2.1.159–61). To prove this point, in a visionary transport to Hades, Lucifer shows Cain a multitude of "phantoms," rational beings "much superior" to humanity, whose "earth is gone for ever—/ So changed by its convulsions, they would not/Be conscious to a single present spot/Of its new scarcely harden'd surface" (*C,* 2.2.69; 2.2.120–23). While Byron assures readers that the existence of prehuman rational beings is "of course, a poetical fiction," his Lucifer also unfolds to Cain images more firmly grounded in the reality of paleontology, exemplifying, for instance, mammoths and perhaps the newly discovered plesiosaurus, "yon immense/Serpent, which rears his dripping mane and vasty/Head ten times higher than the haughtiest cedar/Forth from the abyss" (*C,* 2.2.190–93).

For Cain, rather than inspiring a sublime feeling of awed devotion to God, "The immortal, the unbounded, the omnipotent,/The overpowering mysteries of space—/ The innumerable worlds that were and are" overwhelm him with revulsion at the annihilation of so many previous species and earths (*C,* 3.1.178–80). His geohistorical vertigo culminates in refusal to worship the God responsible for such devastation, in heated argument with Abel, and in the latter's death, instantly regretted by Cain as transforming him into that which he dreaded: a destroyer of the living. While Byron's depiction of deep time led Mary Shelley to view *Cain* as written "in the highest style of imaginative Poetry," conservative moralists, such as Reverend John Styles, condemned the work in words provocatively suggestive of the main disaster in *The Last Man,* calling *Cain* an instance of Byron's "moral pestilence" and "moral disease infinitely worse than plague."[17] Reactions in the manner of Styles redoubled anxiety among geologists not wishing their new science received in Byronic terms. By portraying the Cuvierian destruction of successive species and worlds as Lucifer's means to instigating a murderous brand of atheism, Byron stoked concerns about geology's compatibility with biblical belief and authority.

Yet Byron was not the only one writing geological verses. Some English geologists who followed Cuvier and felt burdened to prove their piety with statements of natural theology in their public works also responded to Byron's drama with poems of their own, exchanging humorous manuscripts within their private circles. William Buckland, for example, an Oxford geologist and friend and correspondent of Cuvier, and whose importance for Mary Shelley will soon become apparent, portrayed himself as a devout challenger of Lucifer in his poem "The Professor's Descent" (1821–22).[18] Dialoguing with Lucifer in a manner unmistakably reminiscent of Byron's Cain, the Professor/Buckland implores Lucifer to impart "some lore of the earth," particularly of prehistoric creatures, such as the plesiosaurus and ichthyosaurus. Instead, Buckland's Lucifer reveals his responsibility for Byron's demonization of geology, responding,

> Mantling in the goblet see
> Boiling sulphur fired by me—
> A drink to madden Byron's brain,
> To nonsense madder still than *Cain;*
> To fire mad Shelly's [*sic*] impious pride
> To final crisis, suicide.[19]

Percy Shelley referenced Cuvier in a note to *Queen Mab* (1813) and possibly in a section of *Prometheus Unbound* (1820).[20] According to Buckland, both Byron and Percy Shelley pollute geology with "nonsense" while being controlled by Lucifer in verse and deed. The geologist willingly accedes to these poets' fate ("D—their souls with all my heart!") but also dismisses them as inconsequential, remaining chiefly concerned with obtaining geological facts, and thus prompts Lucifer to exclaim,

> Ha! no Radical art thou
> Foe of Hell I know thee now.
> Hie thee hence & boast at home
> That never more shall Parson come
> To break my iron sleep again.[21]

Like many geologists of the day, Buckland was a cleric; he later became the Dean of Westminster. Recognizing him as a "Foe of Hell," Lucifer withholds geological secrets that Buckland would use to uphold, not undermine, Christian thought. Buckland's poem thus works to reconcile geology and Christianity, and to bolster the confidence and resolve of his geological colleagues who could here laugh at Byron's expense while repairing the ideological damage caused by *Cain*. Buckland favored humor as a means for geologists to address their self-image out of public view, as in his famously entertaining lectures at Oxford, where he sent students into roars of laughter with his impressions of extinct creatures, a style of teaching that later struck Charles Darwin as buffoonery, but that succeeded in drawing large crowds early in Buckland's career.[22] His jocund approach to geological study helped inspire his students and colleagues to write verses as well, often with Buckland as their subject, and perhaps nothing inspired so much of this doggerel as the cave theory that gained Buckland international renown.

Mary Shelley's visionary glimpses of the future, originating in her findings within the sibyl's cave, capitalize on the cave-mania created by Buckland's work. In December 1821, at almost the same time that Byron's *Cain* went to press, Buckland examined a lately discovered cave in Kirkdale, North Yorkshire. Prompted to investigation by Cuvier, who hoped the cave might yield interesting fossil specimens, Buckland found the cave floor littered with teeth and bones belonging to extinct species of elephant, rhinoceros, hippopotamus, horse, ox, deer, bear, fox, and various birds, with the greatest quantity belonging to hyenas. The large number of hyena remains, and gnawing marks on many of the cave bones, led Buckland to an important ecological theory interpreting the Kirkdale cave as an antediluvian hyenas' den to which the scavengers dragged their prey, sometimes even cannibalizing one another's carcasses. Buckland's theory enabled paleontologists to begin to imagine the living habitats of fossilized species recovered from the earth. In his *Reliquiae Diluvianae* (1823), Buckland attributed various phenomena on the earth's surface to effects of the biblical flood in Genesis, which he integrated with his cave paleontology.[23] Since evidence showed that this cave remained above sea level before the flood as well as in modern times, Buckland imagined that a sudden deluge, perhaps in the form of a giant surge or tidal wave, resulted in extinction of these hyenas and the species on which they preyed.[24] Buckland's theory of hyena ecology vividly brought to life the geohistorical era prior to the modern period

William Conybeare's cartoon of William Buckland entering the cave at Kirkdale in Yorkshire in 1821. (© The Trustees of the Natural History Museum, London)

and inspired the first-known illustrated scene from deep time. In his lithographed cartoon, Buckland's friend and fellow Oxford geologist William Daniel Conybeare depicts Buckland crawling into the Kirkdale cave, where by candlelight the geologist reveals four antediluvian hyenas who appear more shocked to see Buckland than he is to see them. Although both parties find this encounter to be hair-raising, the hyenas' wide-eyed surprise contrasts with the time-traveler's smile of visionary delight as the scene enacts and confirms his geological theories. Buckland's candle symbolizes Enlightenment science's elucidation of the unknown, the geologist's imaginative illumination of antediluvian habitats.[25] The illustration "was probably distributed widely among the members of the Geological Society in Britain; it certainly reached Cuvier, and probably others too, on the Continent."[26]

Conybeare accompanied this illustration with a poem ("On the Hyæna's Den at Kirkdale" [1822]) celebrating Buckland's geological insights and, like Buckland's "Professor's Descent," satirizing Byron. Conybeare's poem describes the hyenas as "munch[ing]" bones "just like Byron's dog," and alludes to their cannibalizing tendencies as a practice to be found in the second canto of *Don Juan*.[27] For Conybeare, Buckland's geological analysis transforms Kirkdale cave into a "Mystic Cavern," a keyhole through which modern humanity can "spy" the distant past, enacting what Thomas Henry Huxley later called "retrospective prophecy."[28] Much of the visionary wonder Byron achieves in *Cain* derives from geology's retrospectively "prophetic" view of the world's past, and of past worlds. His appropriation of this geohistorical capacity profoundly affected Mary Shelley, who declares of *Cain*, "To me it sounds like a revelation—of some works one says—one has thought of such things though one could not have expressed it so well—It is not thus with Cain—One has perhaps stood on the extreme verge of such ideas and from the midst of the darkness which has surrounded us the voice of the Poet now is heard telling a wondrous tale."[29] For Shelley, Byron's "revelation," like Buckland's torch of science and imagination, illuminates the "darkness which has surrounded us" and, through the prophetic powers of geology, "tell[s] a wondrous tale" of what was, is, and will be in the history of the earth. In Cuvier's words, geologists "burst the limits of time."[30] By invoking geology, Shelley enhances her claim to prophetic, time-bursting power her own deciphered revelations, her authorial interpretations, from the sibyl's leaves, found in another "Mystic Cavern."

The cave excursion in Shelley's introduction to her novel recalls Buckland's cave theories. To find the sibyl's cave, Shelley and her companion wind through "murky subterranean passages," groping their way through increasingly narrow and low corridors, until discovering "a wide cavern with an arched dome-like roof" (*L*, 2). Significantly, the faint light by which they find this cavern issues from a fissure in the cave's ceiling. In Buckland's *Reliquiae Diluvianae*, he speculates that some antediluvian cave bones lacking the telltale gnawing marks of the hyena's den belonged to animals that had fallen through fissures in the cave's ceiling before the deluge. With poignant reference to Buckland's theory, Shelley relates that, in the sibyl's cave, "The only sign that life had been here, was the perfect snow-white skeleton of a goat, which had probably not perceived the opening as it grazed on the hill above, and had fallen headlong. Ages perhaps had elapsed since this catastrophe" (*L*, 2). Although the sibyl's historical existence would place this goat's death in postdiluvial times, Shelley nevertheless alludes to an elongated geohistorical perspective, suggesting that "Ages . . . had elapsed," and associating this skeleton with Buckland's fossilized cave bones. Of course Shelley's application of the term "catastrophe" to describe this extinction of past life invokes Cuvier's catchword for his ideas, later known as catastrophism. This image also lays groundwork for understanding the loss of a single life in catastrophic terms, downsizing Cuvierian cataclysm from, say, the shifting of continents to the accidental plummeting of one living being to its death.[31] Fresh in Shelley's memory would have been Byron's exclamation at the cremation ceremony for the sea-worn remains of Percy Shelley and Edward Williams, as she recorded in her journal from Trelawny's account that "Lord Byron looking at the shapeless, limbless mass as it was dragged from out its sandy grave said—'What *is* a *human* body! Why it might be the rotten carcase [*sic*] of a sheep for all I can distinguish!'"[32] For both Byron and Mary Shelley, this essential indistinctness of human and animal remains implies shared susceptibility to extinction. Shortly after Percy Shelley's death, Byron inferred human destiny from paleontological discoveries, citing Cuvier in canto 9 of *Don Juan* to imagine a time "When this world shall be *former,* underground"; he wryly challenges his readers to "Think if then George the Fourth should be dug up!/How the new worldlings of the new East/Will wonder where such animals could sup!"—thus reducing humanity to "such animals" that comprise mere specimens for "a new Museum."[33] Just so, Mary Shelley's transformation of the sibyl's cave into a Buckland-esque bone cave through the goat skeleton from "Ages" past sets the tone for prophesying the extinction awaiting humankind.

The Anticlimax of Natural Disaster

In delineating the plague's extermination of humanity, Shelley may appear to reject Cuvier's catastrophism, his insistence that sudden, violent catastrophes, changing the earth's surface, caused previous extinctions of species; however, Shelley's portrayal upholds Cuvier's ideas for the modern period in which she sets *The Last Man*. Early in his career, Cuvier suggested the possibility of future revolutions of the earth, implicitly forecasting the extinction of present species, including humanity, yet he quietly dropped this radical suggestion from later writings.[34] As his theories evolved, Cuvier decided that the earth's past revolutions, while sudden, had not been universal or global, but rather local and particular to specific regions, becoming more localized and less violent through the course of history. Examining present catalysts of geological alterations on the earth's surface, such as volcanoes and landslides, he declared that in their modern state these forces could not enact the massive revolutions of the past. In her novel, Shelley revives the fearful prospect of human extinction, but does so in a way that corresponds with Cuvier's notion of the earth's diminished power. She demonstrates a relatively moderate destructive potential for current geological and meteorological phenomena, alluding to various theories of the causes and forms of past revolutions and, while making these the companions of pestilence, aligns with Cuvier by denying these forces' responsibility for the future extinction of humanity. Associating her novel with Cuvier's writings and influence, even in the sibyl's cave, Shelley notices the ravages of past revolutions in which "the whole of this land had been so convulsed by earthquake and volcano" (*L*, 3). She portrays natural phenomena of this kind as now largely ineffective in their destructive capacity, and as portents, rather than causes, of the coming extinction.

In the early decades of the nineteenth century, and indeed, even today, geologists' speculations about reasons for past species' extinctions in the Cretaceous period provide no definitive answer. Shelley evokes, only to dismiss, other popular extinction theories in order to assert the primacy of her chosen hypothesis, demonstrated in futurity: plague. For example, with the advent of the plague at the beginning of the novel's second volume, Shelley charts the disease's movement across the earth in an atmosphere of geologic intensity. Originating in the East, the plague quickly spreads within Asia, Eastern Europe, Africa, and into the Americas, coupled with reports of strange natural phenomena and disasters. From Asia and the eastern Mediterranean come accounts that "a black sun" caused a complete eclipse, accompanied by a "convulsion" of the earth "which 'shook lions into civil streets;'— birds, strong-winged eagles, suddenly blinded, fell in the market-places, while owls and bats shewed themselves welcoming the early night" (*L*, 162). Shelley's black orb, which brings absolute darkness, conjures up apocalyptic dread of the extinguishing of the sun, an event predicted in Buffon's theory of cooling planets and the central cataclysm of previous last-of-the-race works by Grainville, Byron, and Campbell. However, for Shelley, this drastic eclipse, and the earth's convulsion, is temporary. She figures the iconographic circumstance of the sun's death as the cause of human (and world) extinction only to avert it, a move that becomes her modus operandi for addressing theories of past extinctions.

In *The Last Man,* Shelley references several theories of the world's past destruction by flood and the event's resonance with both catastrophists and biblical literalists. After the plague reaches England, she dramatizes Cuvier's hypothesis that the earth's last revolution resulted from either flood or a drastic change in temperature, portraying a less violent form of both. Blighting Britons' hopes that cold weather will dissipate the pestilence, plague-stricken England experiences a year without a winter, a striking reversal of the famed "year without a summer" experienced by the Shelleys and Byron at Geneva in 1816. The country also faces a failed crop, devastating storms, and flooding so that "half of England was under water" and, due to such floods on the Continent, "whole villages were carried away" (*L,* 194). Extending this threat of deluge, Shelley sets the stage for Buckland's flood theory, which he envisioned as a sudden surge or tidal wave. In this instance, gathering on Dover's cliffs, the remnants of Britain's population prepare to migrate to the Continent in search of a climate less susceptible to pestilence; namely, they plan to travel to the Alps, which Romantic-era naturalists often described as salubrious to the point of hyperbole—able to cure weakness, cares, and infirmities, regenerate the mind and body, and whose waters "have not crossed the pestilential vapours which hover over our plains."[35] As the soon-to-be emigrants watch the sun setting over an unusually tempestuous sea,

> suddenly, a wonder! three other suns, alike burning and brilliant, rushed from various quarters of the heavens toward the great orb; they whirled round it. . . . The horses broke loose from their stalls in terror—a herd of cattle, panic struck, raced down to the brink of the cliff, and blinded by the light, plunged down with frightful yells in the waves below. The time occupied by the apparition of these meteors was comparatively short; suddenly the three mock suns united in one, and plunged into the sea. A few seconds afterwards, a deafening watery sound came up with awful peal from the spot where they had disappeared. (*L,* 269–70)

With the plunge of these "meteors" into the sea, a "wall of water" rises, prompting the horrified crowd of spectators frantically to question, "Was not the giant wave far higher than the precipice? Would not our little island be deluged by its approach?" (*L,* 270). Yet, as the wave approaches, it dissipates entirely.

By expending the strength of the tidal wave before it reaches land,

Shelley both sustains the scriptural promise against repetition of the flood and rejects the biblically and geologically founded deluge of the past as a possibility for the future destruction of humanity.[36] Through these meteors Shelley both alludes to the flood that, according to numerous geologists, caused the last revolution and extinction of various species, and points to a theory discussed by naturalists such as George Greenough and John Henslow, claiming the flood resulted from a comet's passage close enough to the earth to produce enormous tidal waves that engulfed the continents.[37] Percy Shelley earlier referenced this cometary mechanism in *Prometheus Unbound*, attributing the extinction of "the jagged alligator" and the "earth-convulsing behemoth" to a deluge caused when "some God/Whose throne was in a comet, passed, and cried,/'Be not!' And like my words they were no more."[38] But, as in her other allusions to geological disasters, Mary Shelley posits the possibility of deluge, building suspense and anticipation for an expected apocalypse, only immediately to quell it.

Finally, following the drowning of Adrian and of Raymond's daughter, Clara, Shelley depicts Cuvier's grandest theory of past extinctions. The deaths of these two characters when their boat sinks in a fleeting storm seal the fate of humankind, extinguishing the final hope for producing a new population of humanity, and their drowning of course recollects that of Adrian's real-life counterpart, Percy Shelley. Significantly, after losing these last human companions, Verney dreams "that ocean, breaking its bounds, carried away the fixed continent and deep rooted mountains, together with the streams I loved, the woods, and the flocks—it raged around, with that continued and dreadful roar which had accompanied the last wreck of surviving humanity" (*L*, 325). Verney's dream reenacts Cuvier's radical theory in which continents sank and exchanged places with oceans at various times in the globe's history. In this way, the deaths of Adrian and Clara become bound up with the drowning extinctions of various antediluvian species (including the British hyena). Yet, *as* a dream, Shelley presents Cuvier's grand theory of past catastrophes only to retract this geological possibility in her futuristic setting of humanity's decline.

Shelley's representations of an earthquake, extinguished sun, floods, climate changes, comets, and violent continental shifts, each bring to view a separate scientific hypothesis about the form of cataclysm causing past or possible species extinctions. Nevertheless, for Shelley, none of these natural forces achieves the sustained potency necessary to annihilate humanity, and her portrayal thus, somewhat paradoxically, remains consistent with Cuvier's ideas about geological catastrophes in the modern, enervated state of the world. Instead, Shelley endorses plague, a force repeatedly decimating species within the history of human memory, and an exterminating agent in Thomas Malthus's theory of population, as well as a viable candidate for causing past species extinctions, including that of the dinosaurs.[39] Moreover, in putting forth and then dismissing other naturalists' theories of annihilation, Shelley demonstrates that humanity's extinction can occur without such dramatic displays of geological disaster. Unlike other contemporary last-of-the-race works that envision the end of humanity as inseparable from worldwide apocalypse, Shelley imagines a continuing world where the extinction of humans involves perhaps less disturbance than the disappearance of the plesiosaurus, mammoth, or British hyena before them; in the absence of geological catastrophe, the earth and its animal inhabitants (with the Byronic exception of a single dog) remain placidly indifferent to the loss of human life. In support of Cuvier's directional notion of the earth, Shelley displays that geologists' theories about extinctions of past species cannot apply to the destruction of future populations on a comparable scale. Instead, she shifts catastrophe away from the level of the geological world and into that of the individual.

"And Now a Bubble Burst, and Now a World"

More than the geological and meteorological disasters described by Cuvier, plague, the invisible and mysterious instrument of human destruction, lends itself to focus on the individual. Here, it represents a disaster experienced exclusively by humans and acts insularly, personally on one's body; it brings physical corruption, cataclysm, catastrophe to the individual world, destroying the domestic world of which it forms a part, and revolting (to return to the sentiments expressed by the critic from the *Monthly Review*) with the "minuteness" of its anatomical effects.[40] The character of Perdita exemplifies this notion of the individual's geological significance when, for instance, she pursues the acquisition of knowledge and finds herself to be the greatest "terræ incognitæ, the pathless wilds of a country that had no chart" (*L*, 114). Shelley presents the individual mind especially as analogous and even interchangeable with its surroundings, often referring to the "universe of thought" so that Raymond asserts, "Philosophers have called man a microcosm of nature, and find a reflection in the internal mind for all this machinery visibly at work around us" (*L*, 9, 29, 31, 46). She increases the value allotted to the loss of a single life by using "extinction" in regard both to species and individuals. While Verney often mentions "the extinction of our species" and "the extinct race" of humanity, he also applies the term to domestic units when, for example, he describes "the sturdy labourer . . . weeping over his extinct family," and to individuals when he describes his wife, Idris, by hoping "extinction could not be near [such] a being" and acknowledging that her life "had long been hovering on its extinction" (*L*, 222, 291, 200, 257, 266). This echoes Shelley's appropriation of the term when she wrote of her domestic isolation, of "my companions extinct before me." Extinction becomes a matter of individual loss, as well as the monolithic destruction implied in the extinction of a species.

The domestic values Shelley emphasizes within the familial, communal concerns of the first volume become increasingly important in proportion to their increasing absence in the novel, as death takes its toll on humanity. Indeed, in such crisis, the significance of domesticity gradually overtakes that of the (masculine/public) pursuits of war, commerce, and, as we shall see, even science, in the final volumes. When, before the first appearance of the disease, Adrian returns from aiding Raymond in the Greek wars against the Turks, he remarks the triviality of individual deaths in wartime: "I have learnt in Greece that one man, more or less, is of small import, while human bodies remain to fill up the thinned ranks of the soldiery; and that the identity of an individual may be overlooked, so that the muster roll contain its full numbers" (*L*, 116). Shelley exhibits this pre-plague carelessness of life in the commercial realm as well, where men daily risk death in trade-driven voyages for empire (*L*, 230). Yet, as the plague makes its way around the globe, individual lives take on new value even in these mercenary contexts of trade and war. When the surviving population of America floods back into Britain in what has been recognized as an example of reverse colonization or the empire striking back, Adrian arrests an ensuing battle between these immigrants and the waning British citizenry by drawing the crowd's attention to a single, dying individual, upon which, "every heart, late fiercely bent on universal massacre, now beat anxiously in hope and fear for the fate of this one man" (*L*, 218).[41] In the midst of plague, this individual's death brings cognizance of life's import and reconciles the opposing forces in "a gush of love and deepest amity . . . talking only how one might assist the other" (*L*, 218–19).

In this vein, Shelley provides numerous portraits, some sympathetic, some less so, of lives successively succumbing to the plague and forcing the reader to endure one devastating loss after another. The powerful singularity of death thus materializes, paradoxically, in repeated, brief glimpses of nameless individuals such as the terrified man who drops dead at the cue of a rambling lunatic, the choir member who chillingly expires in the midst of his hymn to God, the old woman who fearfully sequesters herself from her community only to die in search of food, and so on, with the sense of each being representative of countless more, equally individual stories of death. And this demonstrated suddenness with which plague brings destruction once contracted reinforces Shelley's geological analogies of disease.

In debates over the form, cause, and duration of the earth's revolutions, while Cuvier and Buckland claimed the occurrence of sudden and violent catastrophes, biblical literalists questioned what constituted "sudden" considering that, in the Mosaic account, Noah's flood lasted for forty days. In the early nineteenth century, a marginal but growing group of anti-catastrophism geologists began to lengthen the time and lessen the degree of violence necessary to effect the earth's revolutions. Since, in Shelley's novel, the plague destroys humanity over the course of seven years and leaves the earth wholly intact, Fiona J. Stafford views *The Last Man* as anticipating "the uniformitarianism of the subsequent decade."[42] Uniformitarians perceived geological change as cyclical and as occurring over vast periods of time so that the earth remained largely "uniform," thus opposing catastrophists' notions of abrupt and violent past change. However, as I have shown, catastrophist ideas about extinction, cave theories, and natural cataclysms crucially inform Shelley's text. With the lessening magnitude of geological catastrophe, Shelley transfers Cuvier's cataclysmic devastation to the private sphere. If, globally speaking, it took several years for the plague to achieve humanity's extinction, the disease affects the individual very swiftly, and in some cases almost instantaneously. Thus, Shelley shifts the suddenness of catastrophe, as well as the magnitude of its resonance, the successive, geological destruction of previous worlds, described in the works of Cuvier and later Byron, into the successive destruction of *individual* worlds. Just as she applies "extinction" both to individuals and to humanity as a species, so does she alternate this geological thinking about "worlds" in terms of general humanity, describing "the earth" as sick and stating, "All the world has plague" (of course it is humanity, not the earth/world, that has plague), while also recognizing each individual as a world, and each death an apocalypse in the lives of surviving friends and family members (*L*, 170, 175, 180). For Shelley, this latter understanding of catastrophe combats scientific generalizations about extinction that disregard the experiences of individuals.

In *Frankenstein*, Shelley critiques the scientist's tendency toward egotistical absorption in speculations and experiments that make him oblivious to real dangers threatening his domestic world; in her third novel, Shelley models this critique perhaps most clearly through her fictional astronomer, Merrival, whom she compares to Pierre-Simon Laplace, the Romantic era's most renowned mathematician and astronomer. Interestingly, Cuvier dedicated his *Ossemens Fossiles* to Laplace to suggest that the new science of geology would do for conceptions of time what Newton and Laplace had done for space. Shelley notes that Merrival's immense knowledge earns him this prestigious comparison, but portrays him, surrounded by plague, as "far too long sighted in his view of humanity to heed the casualties of the day," regaling any potential listener with "his Essay on the Pericyclical Motions of the Earth's Axis" and "the state of mankind six thousand years hence," for he "lived in the midst of contagion unconscious of its existence" (*L*, 209). Merrival's conjectures, six thousand years into the future, form an exact reversal of those of naturalists and biblical literalists who placed the earth's creation roughly six thousand years in the past; his further speculations about the conditions of the earth a hundred thousand years in the future stands in for much greater estimates of the planet's age by a growing majority of geologists in the early nineteenth century, displaying Shelley's subtle nod to this geological debate, as well as to its frivolity when one's private world is under threat (*L*, 159–60).[43] Unconcernedly impoverished and humorously only half-conscious of his caring wife and boisterous children, Merrival's visionary, telescopic gaze blinds him to present and personal realities that soon break in with painful obtrusion when every member of his family perishes from the disease. Struck with sudden and immense grief, "The old man felt the system of universal nature which he had so long studied and adored, slide from under him, and he stood among the dead, and lifted his voice in curses"; after many days of anguished mourning, he dies while "embracing the sod" of his wife's and children's graves (*L*, 220–21, 221–22). The visionary capacity of astronomers and geologists, capable of grasping the immensities of time, the destruction of former worlds, and the possibilities of the future, becomes meaningless with the destruction of cherished individuals within one's own domestic world. Shelley, as a grieving widow and mother, illustrates that the earth need not undergo a geological revolution for one's world to suffer a cataclysmic convulsion. Detailing the extinction of humankind as a horrifying succession of individual deaths in which no one will survive, Shelley presses the extent to which these deaths devastate domestic circles left behind.

Even the first major demise in the novel, that of Lord Raymond, occurring before the plague crisis, emphasizes death as the catastrophic destruction of an individual "world," represented by Shelley as a literal cataclysm. When fighting for the Greeks, Raymond charges alone into the lately abandoned Turkish stronghold of Constantinople, and suddenly "a crash was heard. Thunderlike it reverberated through the sky, while the air was darkened . . . fragments of buildings whirled above, half seen in smoke, while flames burst out beneath, and continued explosions filled the air with terrific thunders" (L, 144). The gunpowder explosions killing Raymond compose an artificial catastrophe, born out of war. Raymond's death leads to the suicide by drowning of his aptly named wife, Perdita. Perdita's initial reaction to losing Raymond reveals a "revolution . . . in the mind," if not in the earth, when she states, "Look on me as dead; and truly if death be a mere change of state, I am dead" (L, 121, 152). With this interior revolution, Perdita psychologically becomes like the "phantoms" of extinct species from previous "worlds," exhibited by Byron's Lucifer to Cain, who find the present earth unrecognizable. She projects her inward cataclysm onto her geological surroundings, affirming, "This is another world, from that which late I inhabited" (L, 152). With the destruction of her domestic world, Perdita experiences private suffering as global catastrophe.

Shelley's psychologization of geological apocalypse differs from M. H. Abrams's classic argument in *Natural Supernaturalism* (1971), which delineates Romantic writers' "secularization" of New Testament models of apocalypse or revelation.[44] According to Abrams, following the failure of their millennial hopes in the French Revolution, many Romantics absorbed the model of apocalypse into the imagination, producing a revolution of consciousness in which the mind possesses the power to transform the world "into a new heaven and new earth."[45] In Abrams's formulation, Wordsworth's internalization of apocalypse, his imaginative perception "accomplishes nothing less than the 'creation' of a new world," which the poet judges to be something "better" that restores hope and "justifies suffering as the necessary means toward the end of a greater good";[46] yet this tone differs dramatically from that of Mary Shelley's *The Last Man.* Shelley's "new world" envisioned by such mental transformation produces devastation, rather than hope, and lacks this sense of the "greater good," pointing instead toward meaninglessness and death.[47] Her psychologization of catastrophe emphasizes individual death as crisis in its impact on the private sphere as well as on the inward consciousness, enacting a feminine appropriation of this Romantic trope. Arguably critiquing contemporary male poets' turn inward, Mary Shelley demonstrates that this extreme individualism, embedded in her concept of lastness, creates isolation. Her focus on devastation rather than renovation differentiates her "secularization of theological ideas" from that of many Romantic-era male writers so that the model of Paul's conversion experience in the New Testament, making him "a new creature," reborn as though in a recreated world, resonates very differently with Shelley's transformed character Perdita, whose new perception produces not redemption but an agonized feeling of displacement in relation to the outer world.[48]

"New Worldlings"

Perdita's psychological cataclysm, bringing a perspective analogous to that of Byron's extinct species from previous epochs of the earth, incapable of identifying the world in its present form, opens the way to thinking about creatures that may inhabit the globe after human extinction. Resisting this thought of an end to human existence, Verney seeks to convince himself, "Surely death is not death, and humanity is not extinct," hoping for a resurrection of the human population by some means (*L,* 301). Although Cuvier convinced most of his colleagues that fossils represented species no longer extant on the earth, crushing expectations, for instance, of finding mammoths still roaming the American West, some geologists speculated about the possibility of resurrecting into life these past forms. Four years following the publication of *The Last Man,* Charles Lyell, a leading proponent of uniformitarianism, and thus of a cyclical rather than directional history of the earth, looked beyond humanity's future extinction in his *Principles of Geology* (1830) and proposed that ecological conditions may eventually recur to those that gave rise to species of the past. Under such conditions, "Then might those genera of animals return, of which the memorials are preserved in the ancient rocks of our continents. The iguanodon might reappear in the woods, and the ichthyosaur in the sea, while the pterodactyle might flit again through umbrageous groves of tree-ferns."[49]

In her novel, Shelley, too, conjectures about possible beings that may occupy the earth after the extinction of humanity. For instance, when Merrival speaks of humanity's state six thousand years in the future, unaware of the species' proximity to annihilation, he provokes Verney to remark that Merrival "might with equal interest to us . . . describe the unknown and unimaginable lineaments of the creatures, who would then occupy the vacated dwelling of mankind" (*L,* 210). Lyell's imaginings of such "unknown and unimaginable . . . creatures" as resurrected beings of the past struck catastrophists as preposterous, prompting one geologist, Henry De la Beche, to create the cartoon "Awful Changes" (1830) for circulation within the Geological Society in London, taking both the illustration's title and its epigraph, "A change came o'er the spirit of my dream," from Byron's poem "The Dream" (1816). Effectively ridiculing Lyell's concept of a future earth, De la Beche employs the carnivalesque "world upside down" technique of broadsides, popular since the sixteenth century, that evoked such images as a horse riding a jockey, a wife beating her husband, and so on.[50] Here, rather than humans studying the fossils of creatures from the Liassic period, De la Beche portrays a future era in which humanity is "found only in a fossil state." He presents Lyell as "Professor Ichthyosaurus," displaying a human skull and lecturing to other ichthyosaurs and plesiosaurs on the subject of human behavior insofar as it can be inferred from fossil remains. The cartoon's caption humorously renders the tenor of this lecture: "'You will at once perceive,' continued Professor Ichthyosaurus, 'that the skull before us belonged to some of the lower order of animals[;] the teeth are very insignificant[,] the power of the jaws trifling, and altogether it seems wonderful how the creature could have procured food.'" One hears the echo of *Don Juan* as these "new worldlings" examine "the power of [human] jaws" and, in Byron's words, wonder "where such animals could sup!" Despite the frustrations of geologists like Buckland and Conybeare with Byron's poetic license, in an atmosphere of visionary speculation about past and future worlds, geologists found themselves both inspiring and inspired by the imaginative literature of their era.

Shelley's text intervenes between the works of Byron and Lyell to join in these speculations about existence in a post-human world. Upon

Henry De la Beche's cartoon, "Awful Changes" (1830), depicting Charles Lyell as Professor Ichthyosaurus. (© The Trustees of the Natural History Museum, London)

being left as the last man, Verney enters the deserted Italian town of Forli and feels pride in its wide streets and impressive architecture, admitting, "I was pleased with the idea, that, if the earth should be again peopled, we, the lost race, would, in the relics left behind, present no contemptible exhibition of our powers to the newcomers" (*L,* 331). While the species of beings with which Verney imagines the future world to be "peopled" remains unclear, his assertion assumes new meaning in light of geological analogies. Since, in Cuvier's rhetoric, "relics" and monuments interchange with fossilized bones, human anatomy comes into question, and Verney's optimism uncomfortably contrasts with the works of Byron and De la Beche, who imagine the "relics" of humanity appearing not only "contemptible," but also "insignificant" and "trifling" to future "worldlings." Interestingly, Verney's speculations about these "newcomers," the "unknown and unimaginable . . . creatures" who may replace humanity, arguably contain great irony because, for Shelley, he becomes representative of this prospect of post-human existence which also formed the chief subject of her first novel.

In *The Last Man,* Shelley encourages comparison between Verney and Frankenstein's Creature. She makes this parallel strikingly explicit when, for example, the solitary Verney states, "I am a tree rent by lightning," identifying himself with the object that memorably inspired the Creature's creation (*L,* 329). Created from fragments of decaying corpses, both human and animal, the Creature constitutes a new species, and Frankenstein's fear of this species' procreation and ultimate destruction of humanity leads him to abandon construction of the Creature's female companion. Significantly, when Verney, the only human to survive being infected, or perhaps inoculated by the plague, emerges from this disease he feels that he possesses superhuman abilities—abilities almost identical to those of the Creature.[51] Verney states, "My body, late the heavy weight that bound me to the tomb, was exuberant with health; mere common exercises were insufficient for my reviving strength; methought I could emulate the speed of the racehorse, discern through the air objects at a blinding distance, hear the operations of nature in her mute abodes; my senses had become so refined and susceptible after my recovery from mortal disease" (*L,* 250–51). The "mortal disease" from which he "recover[s]" is death and, comparably to both Frankenstein's Creature and the fossilized creatures imaginatively resurrected in the hands of Cuvier, Verney rises from the dead. His survival and transformation renders him, like the Creature, a unique being, perhaps harboring potential for a new biological race or species. But if this is the case, his potential, like that of the Creature, is not to be realized as both figures then represent the first and last of their respective races. In both cases, they are anomalies on the earth. Retaining the resonance of geological catastrophe, these Byronic, solitary figures poignantly individualize the concept of extinction, appearing as temporally alienated, belonging to an extinct (or not-yet-actualized) species, deprived of domestic felicity. For the Creature, his resemblance to humankind only exacerbates his feelings of monstrosity, and Verney echoes those sentiments of displacement, perceiving his person as "a monstrous excrescence of nature" for its likeness to an extinct species (*L,* 340).[52] Upon becoming the last man, Verney even assumes a corpse-like appearance, reminiscent of the Creature: "My tattered dress was that in which I had crawled half alive from the tempestuous sea. My long and tangled hair hung in elf locks on my brow—my dark eyes, now hollow and wild, gleamed from under them—my cheeks were discoloured by the jaundice, which (the effect of misery and neglect) suffused my skin" (*L,* 331). Although Verney tries to revive his hope of finding a fellow being in the world, in the reader's last image of him, as of the Creature, he paddles off in a lone, "tiny bark" (*L,* 342). Despite Shelley's conjectures regarding the generation of future beings, in the case of both Verney and the Creature, as in her positing of modern catastrophic revolutions of the earth, she presents these possibilities only to dismiss them. If future "creatures" will assume the place of humanity, Shelley ensures that their "lineaments" remain "unknown and unimaginable."

This "World's Sole Monument"

Readers seek in vain for some higher purpose or ideological justification for humanity's annihilation in *The Last Man.* Struggling to grasp the irrevocability of extinction, many geologists desired an explanation for such disappearances of the past. In a manner reminiscent of Byron's Cain, Shelley's contemporaries questioned why God would create species only to destroy them. Interestingly, attempting to answer this question, a geologist and friend of Buckland, Philip Duncan, lampooned the "last man" motif with the Kirkdale cave discoveries in his poem "The Last Hyæna" (1820s?; pub. 1869). Portraying a hyena poised on a precipice in sublime solitude just before the flood in Genesis, Duncan's poem combines Romantic aesthetics with Buckland's diluvial theory of catastrophism and bathetically renders the hyena's perspective of his fate. With the remorse of Byron's fratricidal Cain, the cannibalistic hyena appears, to some extent, responsible for his own lastness, having eaten the other hyenas he could find.

> But now the whelming surge had buried all,
> In caves below, of beast both great and small,
> But e'er it rose to mix him with the rest,
> Thus did he growl aloud his last request:
> "My skull to William Buckland I bequeath,"—
> He moaned, and ocean's wave he sank beneath.[53]

Duncan provides a simple explanation for these diluvial extinctions. Satirizing geologists' efforts to reconcile extinctions with the biblical deluge, he portrays Noah's flood as a *felix culpa,* for

> had not man, with deeds of deepest dye,
> Brought down the streaming vengeance from on high,
> And swelled the ocean from its dark retreat,
> His brother monster must have wanted meat.[54]

The British hyena represents only one such "monster" of the many fossilized species recovered by paleontologists in the early nineteenth century whose size and evident or assumed ferocity prompted exclamations of relief at their extinction. Attribution of these species' disappearance as proof of God's wisdom and love for humankind became a commonplace means of conciliating geology and natural theology.[55] But, as Duncan's comical verse implies, most geologists recognized the inadequacy of these efforts toward explanation and reconciliation. Certainly, such an anthropocentric rationalization in which the world functions for the sole benefit of humanity offers no comfort and, indeed, becomes ridiculous, within the context of human extinction on a still-continuing earth. Making no pretext toward some grander scheme or purpose, Shelley depicts nature's utter indifference, writing, "Yes, this is the earth; there is no change—no ruin—no rent made in her verdurous expanse; she continues to wheel round and round, with alternate night and day, through the sky, though man is not her adorner or inhabitant" (L, 334).

No catastrophe marks humankind's disappearance, yet, to return to the puzzling inscription quoted in this chapter's introduction, Verney delineates 2100 as the "last year of the world!" This paradox of the world ending while the earth experiences "no change" and "continues to wheel round and round," suggests that Verney refers to the "world" of humanity, which now signifies only his own life. While not suicidal, Verney alludes to the possibility of his own drowning death as he prepares to set off in his boat on "that water—cause of my woes, perhaps now to be their cure," either in death or in the improbable discovery of other survivors of the plague (L, 341). Shelley's psychologization of geological catastrophism culminates in Verney's prophesied destruction. In his solitude, it is Verney who composes the narrative of Shelley's novel, which she traces from the sibyl's leaves, and which he describes as "this 'world's sole monument'. . . . a monument of the existence of Verney, the Last Man" (L, 339). His quote, "world's sole monument," comes from Edmund Spenser's description of Rome and re-echoes the Cuvierian correlation between "relics" from past civilizations and "relics" or fossils from past species.[56] Verney's characterization of his book as a "monument" thereby recurs to Cuvier's self-designation as an antiquarian, deciphering the "monuments" of extinct species, an employment that transfers to Shelley through her deciphering of this text, this monument that symbolizes Verney's preserved remains. Implicitly, her autobiographical framework further makes this text Shelley's "monument" to her own domestic world, now extinct, yet preserved for future decipherers in the form of savvy readers and literary critics. Just as, for Cuvier, the monuments or fossils of extinct creatures communicate a narrative of their world's past, this monument of Verney's existence details a narrative of his individual history, his own world's past. When Verney carves into stone that 2100 signifies the "last year of the world," he essentially inscribes his own gravestone before embarking on a search for life that can only end in death—in the catastrophic destruction of this particular "world."

Within the transitional moment of the 1820s, when Mary Shelley and many of her contemporaries experienced a sense of limbo following the intense sociopolitical and literary exuberance of the revolutionary and Napoleonic decades, provoking a feeling of posthumous existence, of existing between one era and the next, her psychologization of science interestingly helps signal the decline of natural history's heyday in the literary works of women writers. As illustrated in earlier chapters, women such as Charlotte Smith and Helen Maria Williams experienced late doubts about natural history's ability to impart originality and authority in imaginative literature, demonstrating the difficulty of sustaining such serious, literary-scientific pursuits. Within this context, Shelley engages with natural history somewhat differently. She avoids technical terminology and the appearance of intertextuality, assessing naturalists' scientific assertions less overtly than the previous generation of women writers. Shelley instead highlights science's moral shortcomings in stylistic terms of intense emotion and individualism as well as reaction against the scientific rationalization of nature that are more conventionally associated with the works of, for instance, Percy Shelley and Byron, even as she critiques such constructs of the Romantic tradition. Looking beyond Romanticism, unlike Wordsworth's confident assertion that "nature never did betray the heart that loved her," Mary Shelley's conception of the environment shares more in common with literary images of the subsequent era, such as Tennyson's unsanitized portrayal of nature's red tooth and claw in "In Memoriam," or Robert Browning's Caliban, who describes "extinctions" as a process of meaningless, *random* selection, rather than natural selection.[57]

Anticipating this Victorian literary anxiety about a secular relationship between nature and humanity that leaves the meaning of both in doubt, in Shelley's novel there is no explanation for the annihilation of individuals or of the human species as singled out from the various other species on the planet. However, through the very threat of loss,

Shelley presses the significance and singularity of those individuals and their domestic relationships, representing nature as not in itself interpretable while censuring self-indulgent interiorization as potentially ending in further isolation and catastrophe. She thus conflates a signal trope of Romanticism with the concept of lastness to indicate that, in its extreme, individualism can become monstrous. By placing clearer emphasis on the domestic sphere than on science as the locus of meaning, Shelley exemplifies a pervasive trend in the serious literature of women writers for the next five decades of the Victorian era, as science became, for them, a more volatile source of cultural authority. Yet unlike many of these subsequent women writers, when faced with a growing dissonance between religion and science in which a choice seemingly must be made between the two, Shelley abandons the former in her exploration of the latter. The tendency noted in this chapter, of geologists versifying scientific ideas in manuscript poetry not meant for publication, parallels a larger shift in imaginative poetry as women writers viewed the era's developing, specialized, secularized, and potentially radical ideas about the natural world as less conducive to public verse. My conclusion thus compares Felicia Hemans's published and unpublished geological poetry as exemplifying alterations from earlier ideological possibilities in science to Victorian concerns about feminine decorum that changed women's participation in natural history and encouraged closer stylistic assimilation with conventions that became associated with the Romantic literary tradition of originality.

Conclusion

FELICIA HEMANS, GEOLOGICAL BODIES, AND THE FATE OF ORIGINALITY

Conceptions of literary originality changed during the early decades of the nineteenth century, as did notions of authorship. These transformations occurred in conjunction with key assertions of professionalization in both literature and science through, for instance, lobbying for patent and copyright laws in the 1820s and '30s. At this time, critics increasingly venerated literary ownership and originality that presented itself as unindebted to outside influences, eliding earlier decades' complex negotiations between authorial autonomy and literary borrowings, references, allusions, and plagiarisms. As critical standards more single-mindedly championed independent creation, they became more hostile toward literary works elucidating networks of collection, collaboration, and correction, the very structures often employed by the women writers of my study in displaying their engagements with texts by preceding and contemporary naturalists as well as by imaginative authors.

Alterations in nineteenth-century literary expectations, ruptures between religion and science, a pervading conservatism in British society, and women's exclusion from scientific specialization all contributed to women's literary focus shifting away from science and toward more popular religious and domestic themes as a means to cultural influence as well as literary success. Certainly, following France's Reign of Terror, British society became less tolerant of arguments for equality in gender, race, and class. Concurrently, and in subsequent decades, women and authorship became more complexly confined by scientific theorists who sought to substantiate biological distinctions between the sexes, denigrating women's capacities to practice science and to achieve literary originality. Moreover, religion became increasingly contested as naturalists struggled to reconcile scriptural tradition with new discoveries and hypotheses about the age of the earth and alterations of species. Science's challenges to traditional beliefs combined with industrial progress and lingering devastation from the Napoleonic wars to generate a reading audience that longed for the patriotic stability, continuity, and comfort that writers such as Felicia Hemans portrayed as attainable in the British home.[1]

Hemans's poetry exemplifies and creates changes in women writers' scientific literature of the 1820s and '30s, registering revised principles in both feminine decorum and originality. While her early unpublished geological verses enact the brand of humor and scientific acumen made fashionable by her female progenitors explored in this book, Hemans's published poems on subjects of natural history downplay science, instead stressing familial, religious themes and conforming more closely to the literary values of emotion, imagination, and originality now associated with Romantic convention. Divided into two parts, my conclusion first examines Hemans's opposing models of unpublished and published poetry, revealing transitioning views of science, gender, and literature in the early nineteenth century. Hemans's new paradigm of poetry serves as a case study of the fate of women's scientific literature in the Victorian period, further demonstrated in the second section's broader historical and sociological analysis of this era's alterations in science and literary originality.

Laughter and Lava: Hemans, Feminine Propriety, and Romantic Geology

In 1823 a male critic praised Felicia Hemans's "womanly nature" in contrast with female authors who versified the natural sciences, likening the latter to nondescript natural objects by sneering, "the writers of Natural History make no mention of ['Bluestockings']."[2] For him, scientific-minded women represent an unnatural, unsexed state that receives correction in Hemans, to whose poetry he turns with "pleasure and confidence." Hemans built her literary reputation on cultivating this difference, and she became "the most widely read female poet of the English-speaking world through the nineteenth and into the early twentieth" centuries.[3] Her friend and earliest biographer, Henry F. Chorley, wrote of her, "I never saw [a woman] so exquisitely feminine. . . . Any thing abstract or scientific was unintelligible and distasteful to her."[4] Yet, despite Chorley's and the critic's public distinction of Hemans from earlier women writers through her "distaste" for science, Hemans's knowledge of the natural sciences, in fact, arguably rivals that of those female predecessors, making her choices about when and how to use science important to understanding early nineteenth-century evolutions in women's mergings of literature and natural history.

While poems by such earlier women as Charlotte Smith often exhibited taxonomic acumen in natural history, Hemans's published works represent a different compatibility of science with feminine propriety. To some extent, this shift complies with the familiar trajectory of science's professionalization in the first decades of the nineteenth century, retreating into universities and professional organizations, such as the Linnean Society, the Royal Society, the Geological Society of London, and the British Association for the Advancement of Science (BAAS), which barred women from full participation. As president-elect of the BAAS in 1831, the English geologist William Buckland rationalized excluding women from the society, explaining, "Everybody whom I spoke to on the subject agreed that if the Meeting is to be of Scientific utility, ladies ought not to attend the reading of the papers—especially in a place like Oxford—as it would overturn the thing into a sort of Albemarle [Royal Institution] dilettante meeting instead of a serious philosophical union of working men."[5] Such exclusions from scientific specialization deterred women writers' literary claims to technical knowledge of natural history. Instead, Hemans and other female authors demonstrated closer stylistic assimilation with conventions of originality critically associated with such male Romantic writers as Wordsworth through adherence to less scientific and more domestically oriented themes in their published poetry.

However, the difference between Hemans's published and unpublished verse complicates this perspective of early nineteenth-century science and literature, as well as women's involvement in science. I analyze this disparity especially in two of Hemans's geological poems: "Epitaph on Mr. W—, a Celebrated Mineralogist" (ca. 1814–16) and "The Image in Lava" (1828). Each poem envisions human remains encapsulated within rock but, while the epitaph on the mineralogist remained unpublished during Hemans's lifetime and exhibits a playfully sardonic and intellectual tone reminiscent of earlier women's scientific literature, the later "Image" was published in Hemans's *Records of Woman: With Other Poems* (1828) and assumes a more earnestly feminine appeal to the heart and imagination in the mode of her famed persona, "Mrs. Hemans," denoting a break from previous, technical scientific poetry. While these contrasting poems reveal her geological knowledge and participation in traditions of scientific poetry, they also display her calculated refashioning of that knowledge in accordance with changing expectations of women writers that precluded erudite use of natural history in serious verse, as well as her public pursuit of a different kind of poetic novelty that would influence subsequent women writers.

HEMANS'S UNPUBLISHED, GEOLOGICAL "WILDNESSES"

After Hemans's death, Chorley published for the first time her "Epitaph on . . . a Celebrated Mineralogist" in his biographical *Memorials of Mrs. Hemans* (1836). He explains that the epitaph captures "her livelier humour—the same which in a freak had absolutely made her set one side of a furze-covered Welsh hill on fire, when abroad on a party of pleasure," and discloses, "none, however, of her 'wildnesses' . . . have been published. Many were destroyed as soon as the effervescence of the moment in which they were produced had subsided."[6] In private correspondence, Hemans's sister, Harriet Browne Owen, reveals that the epitaph's mineralogist is C. Pleydell N. Wilton, and that "during one of those 'mountain rambles' so delightfully enlivened by the wit & good humour of Mrs. Hemans," in the midst of searching for geological specimens, "this gentleman unfortunately fell off a rock, while in the act of exclaiming 'Ocular demonstration.'"[7] Wilton, who later held clerical posts in New South Wales and Newcastle, survived his fall and was very much alive to receive Hemans's entertaining epitaphs before embarking for his studies at Cambridge University during the summer of 1816.

In her mock-elegiac poem "Celebrated Mineralogist," Hemans describes rock specimens and geological processes. Imagining Wilton's death, she satirically apotheosizes his mineralogically dedicated life and imbues his tomb with cozy, geological luxuries. The poem's first stanza offers an idea of its humorous tone and content:

Stop, passenger! a wondrous tale to list—
Here lies a famous Mineralogist.
Famous indeed! such traces of his power,

He's left from Penmaenbach to Penmaenmawr,
Such caves, and chasms, and fissures in the rocks,
His works resemble those of earthquake shocks;
And future ages very much may wonder
What mighty giant rent the hills asunder,
Or whether Lucifer himself had ne'er
Gone with his crew to play at foot-ball there. (ll. 1–10)[8]

Hemans's verses on "fossils, flints, and spars," and various other materials and concepts associated with geology, display a tone, topic, and specificity that exhibits continuity with earlier women's scientific poetry but that her posthumous editors, and apparently Hemans herself, now viewed as unbecoming of a serious female poet and thus inappropriate for publication. Her suppression of these verses illustrates her reluctance to associate herself with this movement of women's scientific literature beyond her circle of close friends and family. Nevertheless, her easy engagement with science confirms that this confinement of technical natural history verse to the private realm denotes a conscious choice in shaping her public persona rather than a lack of scientific knowledge. She comically portrays the impact of Wilton's hammer as producing geological cataclysms as he searches for mineralogical specimens, conjuring up Georges Cuvier's catastrophist theories of sudden and violent geological revolutions or changes of the earth so that Wilton's "works resemble those of earthquake shocks" (l. 6). In Hemans's depiction, the mineralogist "hostile[ly]" destroys the earth to gain knowledge of it, yet the expanse "of his power" is relatively small, stretching only "from Penmaenbach to Penmaenmawr."[9] Contemporary geologists, including Cuvier and William Buckland, theorized that the formation of "caves, and chasms, and fissures" resulted from floods, volcanoes, and other natural cataclysms that caused successive destructions of the earth and extinctions of species. Hemans's application of Augustan-style hyperbole, laughably attributing such catastrophes to Wilton's brute strength, dwarfs his "power" through sheer comparison with these natural forces.

Associating Lucifer with geology's underground explorations and revelations of histories of the earth, Hemans anticipates the religious concerns of Byron's *Cain* and Buckland's manuscript poetry, examined in chapter 6 as demonstrating that geology could appear threateningly atheistic in causing humanity to question the purpose of a world filled with devastation and extinction.[10] Even in her jest, presenting Hades as the underground territory of geologists as well as of Lucifer, Hemans's allusion to this satanic potential signifies an audacious move for the female poet. It also contrasts with Wilton's own later poem "Geology," which opens his short volume *Geology and Other Poems* (1818), written in a didactic form that includes long scientific endnotes, piously beginning with an apostrophe to God and the Holy Trinity.[11] By describing Wilton in life as "split[ting] huge cliffs . . . in twain" with his hammer, Hemans converts him into a colossal, "mighty giant" and arguably blasphemous figure akin to "Lucifer," usurping God's power to enact, in this case, geological catastrophes.

In contrast, Hemans's second stanza domesticates geology. She portrays Wilton, this now supposedly lifeless mineralogist, as surrounded by geological equivalents of household comforts, such as fossilized culinary delicacies, so that "Each hand contains a shell-fish petrified:/His mouth a piece of pudding-stone incloses," and his feet are warmed by "a lump of coal" (ll. 20–22). This ludicrous application of the names of natural objects, such as "pudding-stone," is reminiscent of Charlotte Smith's earlier poem "Flora," where, for instance, she employs "Moss Saxifrage, commonly called Ladies' cushion" as the seat for her botanical goddess.[12] By exploiting geological methods' and materials' built-in analogies with food and the comfort of a fire, Hemans creates out of Wilton's tomb the private space of home. However, rather than appropriating geology as a feminine space, she here subordinates domestic allusions to those of science, enhancing the verses' playful tone; indeed, for the mineralogist, science itself symbolizes home and replaces that domestic function. In this way, Hemans's depiction could be read as a humorous verse equivalent of Mary Shelley's critique of the solitary, obsessive Victor Frankenstein, who casts aside genuine domestic comforts for those of science. As Susan Wolfson notes, Hemans's poem is "peppered with nearly hudibrastic rhymes—and parodies of elegiac conventions and tropes."[13] These qualities echo earlier women writers' incorporations of such neoclassical ("mock") styles of poetry when portraying science, aligning Hemans's unpublished verses with this tradition of women's scientific literature.

In the poem's final stanza, Hemans modestly and ironically disclaims geological acumen, undercutting her naming of particular kinds of rock (such as schist and gneiss) with statements of "whate'er ye be" and "names too hard for me" while displaying that she does possess this scientific information (ll. 29–30). Her self-deprecating gestures within this satire work both to include and exclude Hemans from knowledge of geological studies, for, to some extent, one must be *in* on the joke to *make* the joke. Just as Barbauld in her poem "The Invitation" transforms naturalists into natural objects for her own study, Hemans likens the expired Wilton to a mineralogical specimen for future discovery and analysis. By sardonically observing male observers of nature, these female poets turn the tables on a scientific discourse that often conflates *women* with natural objects. In the end, Hemans depicts Wilton as entombed within the very rocks he once sought and studied, a buried fossil himself, "as cold as any of his petrifactions" (l. 36). Hemans's tongue-in-cheek style allows her, as it did many women writers of the previous generation, to assume a removed and superior sense of scientific participation, frankly capable of enjoying natural history's lighthearted as well as more studious aspects. However, Hemans withheld from public view this humorous poetic treatment of scientific subjects and, perhaps less expectedly, anticipated verse techniques practiced by geologists.

"A MONUMENT RAISED TO HIMSELF": GEOLOGISTS' POETRY

Hemans's epitaph exemplifies a mode of satirical, scientific poetry that became popular among geologists in the early nineteenth century. As Ralph O'Connor has shown, English geologists such as William Buckland sometimes wrote verses as an aspirational genre, exploring how they wished themselves and their developing discipline perceived.[14] However, like Hemans, in not publishing her geological "wildnesses," these geologists also generally did not publish their experimental jocular verses, instead circulating them in manuscript or in privately printed broadsheets. This suggests that the withholding from publication of such poems was not solely a matter of Hemans's gender, but that it also signifies pervasive concerns about religion, tone, aesthetic taste, and professional personality in the first quarter of the nineteenth century. Although there is no evidence that the Oxford geologists such as Buckland were aware of Hemans's epitaphs, her unpublished verses forecast certain trends found in their poetry. Unwilling to let these naturalists' poems go unknown, Charles Daubeny, professor of chemistry and of botany at Oxford, collected together a number of geologists' manuscript poems, spanning from the 1820s through the 1860s for publication in a volume titled *Fugitive Poems connected with Natural History and Physical Science* (1869). This collection includes lampooning verses about, for instance, mastodons, trilobites, fossil caves, "the origin of species," and "the fate of the dodo." With striking similarity to Hemans's epitaphs on Wilton and on his hammer, *Fugitive Poems* contains comical verses about Buckland, such as William Conybeare's "Ode to a Professor's Hammer" and Richard Whately's "Elegy Intended for Professor Buckland," dated December 1820 and thus thirty-six years before Buckland's actual death in 1856.

Whately's elegy on Buckland, written at least four years after Hemans's epitaph on Wilton, envisions its defunct geologist as a geological specimen. Predicting a different fate for Buckland than Hemans earlier imagined for her mineralogist friend, Whately wonders:

> Where shall we our great Professor inter,
> That in peace may rest his bones?
> If we hew him a rocky sepulcher,
> He'll rise and break the stones.
> And examine each stratum that lies around—
> For he is quite in his element underground. . . .
>
> Then exposed to the drip of some case-hardening spring,
> His carcase let stalactite cover,
> And to Oxford the petrified sage let us bring,
> When he is incrusted all over;
> There, 'mid mammoths and crocodiles, high on a shelf,
> Let him stand as a monument raised to himself.[15]

While Hemans's Wilton rests comfortably in death when encompassed by his mineralogical finds, Whately describes Buckland's corpse as restless in such a geologically rich environment. Unlike Hemans's simile describing Wilton, "*as* cold *as* any of his petrifactions" (emphasis mine), Whately literally "petrifie[s]" Buckland's corpse as the single means to prevent it from reviving. Eager even in death to examine his geological surroundings, Buckland becomes incapacitated only when calcified into a "stalactite," forever standing erect in the figure of an eerie "monument raised to himself." Moreover, rather than placing this statue-Buckland in a location of public viewing or reverence, Whately imagines it sequestered and catalogued "high on a shelf" as merely another fossilized specimen, kept simply for the sake of collection, and all but forgotten. Whately, like other geologists of his day, humorously displays a paradoxical confidence in his discipline, portraying this science, as well as one of its best-known thinkers, as simultaneously everlasting and negligible. Conflating Buckland with the many stalactite-encased extinct species he unearthed, Whately symbolizes professional progress as both inconsequential and unstoppable except through a sort of self-destruction, a human extinction through geological processes. Just as Hemans kept her satirical scientific verses private, Whately and other Oxford geologists explored these potentially controversial subjects out of public view, instead publishing more serious presentations of their ideas and authorial personae. In her later poem "The Image in Lava," Hemans describes another rock-enclosed "carcase," a victim of geological disaster, that raises similar questions about its relative importance and duration while embodying a more solemn "monument" than this portrayal of Buckland, and an alternative mode of originality in women's scientific poetry.

"PRINT UPON THE DUST": A NATURAL HISTORY OF WOMAN'S LOVE

Perhaps the most crucial distinction separating Hemans's "Image in Lava" from her earlier epitaphs and from poems by geologists such as Whately and Conybeare is that she intended these verses for publication. Included in *Records of Woman* (1828), and thus published more than a decade after she wrote the geological epitaphs, her later poem approaches science in poetry differently, adopting a more serious, elevated, and feminine tone. As Hemans's note to the poem explains, its title and content refer to "the impression of a woman's form, with an infant clasped to the bosom, found at the uncovering of Herculaneum," a city that, along with Pompeii, was buried in the "sudden catastrophic" volcanic eruption of Mount Vesuvius in AD 79. Surprised by this disaster, many of these cities' inhabitants did not have time to escape from the enveloping toxic gas or the pumice and volcanic ash that rained down from the eruption, searing skin and asphyxiating these unfortunate victims. Among those killed was the famous naturalist Pliny the Elder, who ventured too close in pursuit of scientific observations.[16] Herculaneum was not discovered until 1709, "and Pompeii in 1748, with major excavations from 1763 to 1820."[17] Over time, some buried human frames left hollow impressions in the hardened ash and pumice as these bodies decomposed. Interestingly, although Hemans places this "impression of a woman's form, with an infant" in Herculaneum, current scientific evidence suggests that this mother and child more likely resided in Pompeii. A recent study states that "the people of Herculaneum did not die from asphyxiation, as in Pompeii, but due to rapid exposure to the intense heat that developed from . . . the immediate boiling and vaporization of their organic liquids and tissues. . . . This is why in Herculaneum the layer of ash did not reveal cavities corresponding to the bodies of victims, a situation that in Pompeii made it possible to reconstruct figures by pouring plaster into spaces that had been left by the progressive disintegration of human tissue around the skeletons."[18] It therefore seems appropriate to identify Hemans's "image" as a figure from Pompeii, rather than from Herculaneum as she claimed. Indeed, a specific female victim discovered in Pompeii may have inspired Hemans's description, for "in 1812, along the Via delle Tombe, near the Villa of Diomedes, the body of a young, bejeweled woman was found, hugging her small child to her chest."[19] Hemans likely knew of such "images" by reading accounts of contemporary excavations. Isobel Armstrong, for instance, explains that "in 1827 *The Times* of July 4 and July 16 carried reports of new finds at Pompeii."[20] In the 1860s excavators began creating plaster casts from the site's impressions of human forms, but of course these did not exist when Hemans published her poem in 1828, and she describes the "image" simply as a "seal," a "print upon the dust."[21]

Hemans's poem depends on her readers' basic knowledge of the pyroclastic processes that created this "image in lava," in the ash and pumice rock resulting from lava flows. Yet her treatment of geology here is much less specific and erudite than in her earlier, unpublished works. Hemans's depiction of a catastrophic destruction that preserves in ashes the tender relation between mother and child dominates the verse, subordinating scientific concerns to those of domesticity. Through its focus on this domestic relationship, Hemans's poem brings to life this woman and child, figures whom history traditionally overlooks. Ironically, their destruction is also these two individuals' survival, for the rock in which they are encapsulated preserves them, as well as their culture, for future analysis and understanding. Unlike Whately's embodied "monument" of Buckland that, "high on a shelf," acquires little significance after the geologist's death, Hemans's "image" invigorates tropes of immortality in both nature and art. She celebrates lived domestic history, and less directly addresses the geological details of this immortalization.

Indeed, Hemans constructs an underlying geological framework to value this domestic relationship between mother and child above the

Skeleton of a woman found at Herculaneum, known as the "Ring Lady" because of the jewelry found with her. The effects of intense heat vaporized her organic tissues, producing no body cavity in decomposition, in contrast with Hemans's description and the Pompeiian pair in the figure pictured on the next page. (O. Louis Mazzatenta/National Geographic Creative)

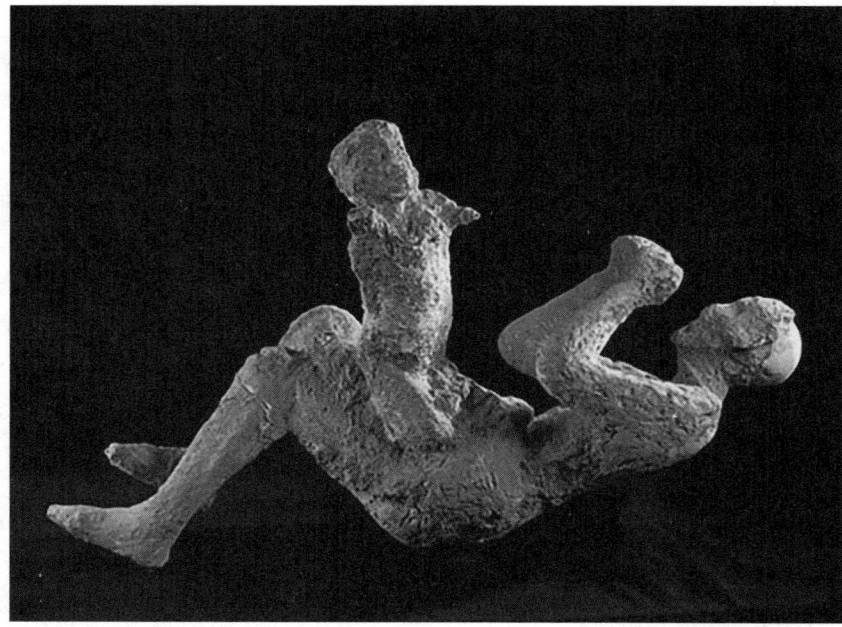

Plaster cast made from the impression of body cavities of a mother and child found in Pompeii. (© Jackie and Bob Dunn, http://www.pompeiiinpictures .com; by permission of the Ministry of Cultural Heritage and Activities and Tourism—Superintendency of Pompei)

masculine realms of science and national affairs. She employs Georges Cuvier's comparison between remains of past civilizations and those of past biological species, in the form of fossils, as discussed in chapter 5. In his analogy, Cuvier applied antiquarian terms, such as "relics" and "monuments," to depict his paleontological discoveries. Hemans arguably references Cuvier's theory of catastrophism when she describes Vesuvius's eruption as "sudden" and "catastrophic" in her footnote to the poem, and appropriates his correlation between science and antiquarianism, writing of the "image":

> Oh! I could pass all *relics*
> Left by the pomps of old,
> To gaze on this rude *monument*,
> Cast in affection's mould. (ll. 33–36, emphasis mine)[22]

Describing this image in lava as a Cuvierian fossil or "monument," the poet subtly situates her poem's subject within geological discourse only to then make science and other masculine institutions less important in comparison with the feminine realms of domesticity and art.

From the poem's beginning, Hemans sets up a gender division, associating the private realm with "woman's heart," capable of outlasting the public and masculine fame of "proud memorials rear'd / By conquerors of mankind" (ll. 7, 11–12). Rather than being intangible and ephemeral, love emerges as more permanent than the monuments of empires and civilizations. Preserved by nature, this "image" transcends culture and appropriates for the feminine an ability to represent the "human" more generally in "Love, human love!" Hemans draws attention to the image's significance as a natural object that becomes, in a sense, artificial, becomes art, and a symbol of that transcendent "heart," speaking across particularities of time and space.

Critics such as Julie Melnyk have explored Byron's early influence over Hemans's poetry, as well as Hemans's later exchange of Byron for Wordsworth as her poetic model.[23] While Byron's satirically geological *Cain* influenced Mary Shelley's scientific writings, Hemans, learning of Byron's personal improprieties and of his scathing remarks on herself and on her verse in Moore's biography of Byron, turned to Wordsworth as displaying a compatibility between poetry and the domestic sphere that she could further develop in her own works. Indeed, as Angela Leighton remarks, Hemans "ventriloquizes the work of her most admired contemporaries and reproduces it with technical efficiency," while constantly subjecting "the imaginative landscapes of Romanticism . . . to the critical and social bias of the woman."[24]

In many ways, Hemans's published poetry compares more readily with that of her male contemporaries than with her earlier geological epitaphs and the movement of women's scientific literature. As other scholars have noted, her ekphrastic "image in lava" harbors particularly pertinent similarities with Keats's "Ode on a Grecian Urn" (1819).[25] Although Hemans expresses a dichotomy between the ruins of empire and the permanence of domestic affection, she imagines the possibility of ruin creeping into the private sphere as well: the collapse of this relationship between mother and child that would occur if the child lived to maturity. By becoming the "image," suspended in time, in this sexually charged "moment of a thousand pangs," the relationship's painful separation will never occur. As on Keats's urn, there is no growing old, no possibility of future change and disappointment. Yet, unlike the static moment of Keats's urn that endlessly defers the lovers' passionate consummation, here mother and child embody the impassioned moment forever. Rather than eternal longing and anticipation, it is an eternal realization—an eternal ecstasy and "agony" of maternal love. In their death and preservation, the "image" arguably fulfills a mother's selfish wish to keep the child young, loving, and dependent, always. Hemans's two human figures, forming a single image, simultaneously are self-contained as one another's entire world and model an alternative, transcendent ideal, a lasting symbol of intensity and purity in attained "affection."

Hemans accentuates this symbolism in the poem's final stanza where, in the phrase "Thou art," "art" functions both as a form of the verb "to be" and also, of course, as the art of the artist (l. 42). She employs the idea of art's immortality, but instead of an outside laboring artist being remembered through his work, the creation and creator are fused. The woman both holds her "art," her child, in her arms and is her own message to posterity; she is her own immortality. When, in the poem's final line, Hemans presses, "It must, it *must* be so!" she affirms the power and permanence of woman's affection, of human love, and yet the repetition betrays doubt, an effort to convince herself of love's efficacy that undermines this authoritative claim (l. 44). Her identification of art with this relief of human forms indicates their interchangeability in this "print upon the dust" (l. 38). While asserting everlasting presence and influence, this symbolic "print," her verse, preserved only in "dust," is also ephemeral, "fragile," and easily lost, supporting her subtle uncertainty.[26] She casts this print's power as "holiness," as spiritual rather than political or scientific, and thus more readily connected with the home (l. 43). The poem's immortalization of "woman's heart" displays the feminine propriety audiences associated with the work of "Mrs. Hemans," and its prominent domesticity overwhelms the verses' obvious but unexplored scientific potential (l. 7).

Differences between Hemans's unpublished and published geological verses illustrate significant changes in early nineteenth-century women writers' integration of natural history and literature, especially serious poetry. According to Leighton, Hemans "accepts rather than rebels against the limitations of her gender" and becomes "one of the true originators of a line of poetry" that, through "the exclusion of money, sex, power and, as it were, imaginative *in*sensibility from the poetic consciousness of women," manifests "a more general, moral protection campaign of Victorian womanhood."[27] Science may be added to the list of pursuits Hemans largely precludes from her published works in establishing this new form of women's literary originality. Her boldly playful scientific unpublished epitaphs markedly differ from the serious and feminine "Image in Lava," which she meant for a public audience, contributing to Victorian ideals of feminine propriety, with emphasis on religion and the domestic sphere, that would influence subsequent female writers such as L.E.L., Christina Rossetti, and Elizabeth Barrett Browning. This shift largely occurs because, by the late 1820s and '30s in science, women (and the public more generally) are no longer supposed to be *in* on the joke, are more pronouncedly excluded from the explorations, discoveries, and self-determination that shaped perceptions of geology and other developing disciplines of natural history. While both Hemans and geologists tested science's playful potential in private, they presented very disparate, more earnest and calculated professional identities to the popular imagination.

Rocks and Religion: Hemans, Nature, and the Wordsworthian "Spirit"

In her published poetry, Hemans's use of science resembles that of Wordsworth more than that of Charlotte Smith or Erasmus Darwin. Geological references in the published verses of Wordsworth and of Hemans rely upon metaphor and imagination, not the learned exploration of their scientific subject that each displays elsewhere. Just as Hemans exhibited geological knowledge in her unpublished verses, "instances of Wordsworth's knowledge of the names and descriptions of rocks and their constituent minerals are infrequent" in his poetry, but fully demonstrated in his prose topographic description of the Lake District, *A Guide to the Lakes* (1810).[28] One of Wordsworth's most often cited poetic allusions to geology occurs in "Resolution and Independence" (1807), where the speaker describes the old Leech-gatherer as yet another geological body:

> As a huge stone is sometimes seen to lie
> Couched on the bald top of an eminence;
> Wonder to all who do the same espy,
> By what means it could thither come, and whence;
> So that it seems a thing endued with sense:
> Like a sea-beast crawled forth, that on a shelf
> Of rock or sand reposeth, there to sun itself. (ll. 57–63)[29]

In their respective scholarship, Alan Bewell and Noah Heringman each analyze this stanza's reference to glacial erratics, explaining the contemporary opinion that only the biblical Deluge could transport non-native boulders, as well as foreign, now-fossilized marine and fauna remains to inland locations.[30] Bewell portrays the sea-beast as transforming into a fossil, a mediating image of sentiency that correlates the Leech-gatherer with rock.[31] By describing "this Man; not all alive nor dead" as an antediluvian "stone" or living fossil, Wordsworth conjures up his era's geological theories of catastrophism as well as the indefinite expansion of time into the distant past, depicting the old man's body, "bent double, feet and head / coming together in life's pilgrimage," like the serpent with its tail in its mouth, symbolizing eternity (ll. 66–67). In these geological allusions, Wordsworth eschews technical language and scientific footnotes while aesthetically appropriating science's inspirations of "wonder" through seemingly ungraspable events and temporalities (l. 59).

As in Hemans's later "Image in Lava," Wordsworth's depiction collapses the boundaries between humanity, nature, and art, as well as between life and death, public and domestic. According to a young Wordsworth in *The Ruined Cottage* (1797–98), natural history's attention to minute details, accurate description, and classification threatened both imaginative poetry and humanity's potential for sensation and sympathy, for "'solitary objects . . . beheld / In disconnection' are 'dead and spiritless', and division, breaking down 'all grandeur' into successive 'littleness', is opposed to man's proper spiritual condition, in which 'All things shall live in us and we shall live / In all things that surround us.'"[32] Taxonomic division has no place in this spiritual communion. Later in life, Wordsworth's more orthodox spirituality caused him to reevaluate success in scientific pursuits. In 1829 he delineated two separate categories of scientists, disapproving of naturalists concerned with "a bare collection of facts for their own sake, or to be applied merely to the material uses of life," while "venerat[ing]" those naturalists who bend their work toward "elevating the mind of God."[33]

Especially as his spirituality became increasingly conventional, Wordsworth represented to Hemans a mentor whose poetry modeled religious interactions with nature that she could then alter for her own purposes.[34] For both his personal and poetic ease within the domestic sphere, she called him "the true *Poet of Home*."[35] As Melnyk argues, Hemans's admiration for the older poet "allowed her to recuperate the vatic poetry of his early maturity for a new age and for a woman poet by extending and expanding the tendencies of Wordsworth's own poetic development toward a more Christian and even more domestic vision."[36] For example, Hemans overtly adopts and adapts Wordsworth's versifications of the natural world in her poem "Wood Walk and Hymn" (1834), employing the closing lines of Wordsworth's "Nutting" as its epigraph:

> Move along these shades
> In gentleness of heart: with gentle hand
> Touch—for there is a spirit in the woods.[37]

Hemans's work comprises a dialogue between a father and his child walking through the forest, naming and sometimes describing a number of trees and plants, including aspens, chestnuts, eglantines, violets, cowslips, woodbines, arums, passion flowers, wood hyacinths, and anemones. Unlike her female predecessors' didactically scientific goals in similar dialogues written to impart botanical nomenclature and anatomy, Hemans's primary aim is to Christianize Wordsworth's "spirit in the woods." In doing so, she employs a form of natural theology, inspiring thoughts of God through observations of nature. However, instead of science, Hemans cites "the peasant's legend," sources of folklore affiliating these plants with Christianity. For instance, Father explains, "rustic[s]" attribute the aspen's "for ever trembling" leaves, its "restlessness" to their notion that Christ's cross "was framed of aspen wood" (ll. 9, 10, 13). Acknowledging the mythical quality of such antiquated ideas by stating, "*We* walk in clearer light," Father nevertheless admits "a lingering love" of reading these religious "characters" "on rock, or herb, and flower" (ll. 19–20, 21, 22, 28). Hemans's poem thereby highlights plants' Christian "characters," rather than the characters that enable naturalists' classifications of those natural objects.

When this parent and child reach "the very inmost heart/Of the old wood," Father contrasts "the days/Of pagan visions" with present knowledge of "our God, a Spirit; who requires/Heart-worship, given in spirit and in truth" (ll. 74–75, 81–82, 90–92). In obvious reply to the ending lines of Wordsworth's "Nutting," the child in Hemans's poem recites a "Wood Hymn," affirming, "Yes. . . . /There *is* a power, a presence in the woods," for "Thou, *thou* art here, my God!," thus reframing Wordsworth's conclusion in this Christian context (ll. 128–29, 133). In her published verses, Hemans, like Wordsworth, avoids serious scientific participation and, indeed, spiritualizes or Christianizes nature at a time when religion became increasingly incompatible with science. While didactic scientific verses like those of the era's earlier women writers and of Erasmus Darwin lost popularity, her poetry acquired an extensive audience and represented reconfiguring notions of originality.

DIVINE ORIGINALITY AND DARWINIAN COMPLEXITY: GENDER AND PROFESSIONALIZATION

In literature and science throughout the nineteenth century, originality and gender constituted interconnected and contested concepts. As Patricia Murphy has shown, numerous nineteenth-century essays and treatises composed by biologists, physicians, anthropologists, psychologists, and other scientists sought to provide "objective" and conclusive evidence to support "long-standing cultural perceptions of an innate female inferiority."[38] The capacity for originality became a central focus of this supposed sexual distinction, portraying men as excelling in complex abstractions, judgment, and novelty while restricting women to greater sensory ability, perception, and rapidity of thought, thereby repeating the dichotomy against which Wollstonecraft earlier warned.[39] The zoologist W. K. Brooks wrote, "The originating or progressive power of the male mind is shown in its highest forms by the ability to pursue original trains of abstract thought, to reach the great generalizations of science, and to give rise to the new creations of poetry and art."[40] For Brooks, as well as for Charles Darwin in his *Descent of Man, and Selection in Relation to Sex* (1871), women aligned with categories of intuition, instinct, imitation, and a "feminine passivity" rather than with the masculine "originating element."[41] In 1865 T. H. Huxley stated, "Women are, by nature, more excitable than men—prone to be swept by tides of emotion," and the contemporary anthropologist Luke Owen Pike echoed, "If man's highest prerogative is to think, women's noblest function is to love."[42] Perceiving this scientific trend as pervasive within Victorian society, Christine Battersby argues in *Gender and Genius: Toward a Feminist Aesthetic* that "women—seen primarily as sensitive, emotional, passive, intuitive, and imitative—were believed capable only of transmitting or nurturing genius in males as wives, mothers, daughters, and sisters."[43] Thus, women often authorized their entry into the literary profession as an extension of their feminine duties. As we have seen, Hemans, for example, achieved professional success through feminizing and Christianizing themes "ventriloquized" from male contemporaries, thereby making them "new"; this establishes a different kind of novelty through selective imitation, correction, and improvement from that defining the scientific collection and collaboration of earlier women writers, displaying her authorial confinement within developing standards of gender and originality. The general Victorian attribution of originality to men retrospectively helped enshrine male poets' reputations of autonomous creativity in the Romantic era, a perspective that became entrenched as early as the 1820s and '30s.

In the latter half of the eighteenth and first half of the nineteenth centuries, imaginative writers and their critics increasingly disparaged neoclassical conventions of imitation while reverencing the notion of literary autonomy. Seeking literary originality through science, the female authors of my study often referenced scientific as well as literary texts, creating a framework of collaboration, allusion, and borrowing that enabled them to elaborate on and contribute ideas in pursuit of scientific accuracy and imaginative novelty. However, reliant on such blatantly referential structures of composition, in the early decades of the nineteenth century these authors' scientific approach to the quest for originality became relegated to secondary literature. Debates within major literary periodicals in the 1830s widely deplored accounts of originality associated with collection, appropriation, or derivation, championing instead creation ex nihilo, or out of nothing. As John A. Heráud stated in an 1830 article of *Fraser's Magazine,* "What are called original thoughts . . . are underived, indeed original existent in the individual soul . . . the imagination creates its ideas . . . from nothing!"[44] Earlier, in his 1818 lectures, William Hazlitt famously portrayed originality as the great aspiration of the "Lake school" of poetry, comprising "new and original" compositions.[45] Members of this first generation of male Romantic poets also encouraged perceptions of their works' unindebted originality, as when Coleridge wrote of Wordsworth in *Biographia Literaria* (1817), "in imaginative power he stands nearest of all modern writers to Shakespeare and Milton; and yet in a kind perfectly unborrowed and his own."[46] Such critical and authorial standards of autonomous originality contributed to literary elitism in the developing profession of letters.

By the 1830s print culture and particularly periodicals had become perhaps the first literary field to acknowledge authorship as a profession.[47] This professionalization further established critical distinctions between literary genres and opened new possibilities for gendered authorship at the same time that it obscured women's past literary successes. Harriet Martineau, for example, the most prominent woman of letters of the 1830s, and possibly of the century, presented herself as a "professional son" and "citizen of the world," a "solitary young authoress" with "no pioneer in her literary path," thus voicing the same career aspirations and ideas of professional authorship as many of her male contemporaries, but in language that arguably ignores the literary accomplishments of female predecessors.[48] Concurrently, as Clare Pettitt recounts, inventors and scientists vigorously campaigned in the 1830s and '40s for new patent laws to secure intellectual property and middle-class incomes.[49] Frank Turner traces how young lay scientists sought professionalization by reclaiming power from clergyman-naturalists (who had so vitally and prominently encouraged science's collective, collaborative pursuits in previous decades) and attaining permanent positions of influence within universities.[50] Endeavors toward professionalization within the sciences thus coincided with literary authors' public discussions of professional income and the introduction of domestic and international copyright bills in Parliament from 1837 onward.[51] Indeed, on the domestic front, Wordsworth, Southey, and Carlyle each publicly urged perpetual copyright laws at this time.

Such coinciding efforts to secure professionalization and individual recognition in science and in literature paradoxically worked to reinforce the fields' separation. In British literature, Robert Macfarlane views 1840 "as the high-water mark" for conceptualizing originality as owing solely to the author, rather than to outside sources, due to key texts by John Stuart Mill and Thomas Carlyle in that year.[52] As M. H. Abrams explains of Mill's essay on Coleridge, "in so far as a literary product simply imitates objects, it is not poetry at all. As a result, reference of poetry to the external universe disappears from Mill's theory, except to the extent that sensible objects may serve as a stimulus or 'occasion for the generation of poetry.'"[53] Denigrating verse that "imitates objects" or examines the "external universe," Mill denies the status of poetry to the previously popular scientific literature examined in this study, reliant on close observation and imitation of natural objects. Similarly, in his lecture "The Hero as Prophet," Carlyle isolates the original genius, invoking Romantic-era associations between the poet and God in the ability to create out of nothing so that "the words he utters are as no man's words" and thus do not develop from description, influence, imitation, or repetition, but aesthetic and intellectual autonomy.[54] While the study

of God in nature helped vindicate women's earlier scientific compositions, the increasing secularization of science now made incompatible
divinely inspired originality and cutting-edge, scientific authority.[55] As
the works of Mary Shelley and Felicia Hemans exemplified, geological discoveries challenging scriptural traditions of time and origins diminished women writers' ability to justify scientific analysis through
natural-theological aims so that these authors distance themselves either
from theological perspectives, as is the case with Shelley, or from technical science, as Hemans does in her published poetry, instead turning
toward religious themes.

Of course this is not to say that Hemans and the succeeding generation of women writers ceased referencing the natural sciences in their
published poetry. As I have shown, Hemans's "Image" indirectly relies
on geology, and she elsewhere adheres to conventions of natural theology, as in her volume of verse *Hymns on the Works of Nature* (1827). In Letitia Landon's poetic text *Flowers of Loveliness* (1838), which may be represented by "The Night-Blowing Convulvus," the identification between
flowers and women reinforces a pervasive feeling of being trapped, op-
pressed, earthbound, and claustrophobic, so that women's condition
reflects plants' rootedness to the spot. She uses the flower as a starting
point, and then transcends it, not to soar to sublime and philosophical
heights like male contemporaries such as Wordsworth, but to elucidate women's domestic trials and *inability* to transcend their material circumstances. Christina Rossetti's "Goblin Market" (1862) also displays
well-known interest in natural history as the goblin men represent zoological species from various colonies within the British Empire. When
Laura cries over the barren seed of her exotic fruit, Rossetti arguably
gestures toward botanists' difficulty sustaining, even in hothouses, valuable plants brought back from these tropical locations.[56] Cautioning
about the global aims of natural history and imperialism, Rossetti ultimately urges that Britons, and particularly women, turn their attention toward aiding one another in charitable domestic goals. However,
these women's published verses on nature do not apply technical terms
of science or strive to generate new information, theories, or discoveries
in natural history as had the works of some women writers of the previous generation.

Notably displaying vicissitudes in literary taste that conjoin with
changes in science, although critical standards of the 1820s and '30s elevated autonomous authorship and denigrated the collaborative forms

of novelty in the scientific literature I have discussed in this book, the decades following this late-Romantic simplification and mythification of originality, the 1840s and '50s, saw a reaction in which imaginative writers embraced literary allusions, borrowings, and intertextuality; this shift coincided with the publications of such scientific texts as Charles Lyell's *Principles of Geology* (1830), Robert Chambers's *Vestiges of the Natural History of Creation* (1844), and Charles Darwin's *On the Origin of Species* (1859). Amid these evolutionary theories that repetition with variation over time could produce something new, eroding past adherences to species fixity, occurs, in turn, a growing devaluation of literary autogeneity. Standards of composition move toward what George Eliot termed "communistic principles" that "treat the distinction between Mine and Thine in original authorship as egoistic, narrowing, and low."[57] This later era's greater acceptance of complexity when thinking about original forms acknowledges an indebtedness to ancestral lineage in both the biological sciences and in literary thought, with recognition of interconnection across deepening notions of time.

Nevertheless, with the growing professionalization of both science and literature, imaginative authors of the Victorian period generally avoided claims to new scientific contributions or discoveries of their own, in contrast to this earlier generation of women writers that interacted with and challenged naturalists while showcasing their personal observations. Natural history books and essays continued to be bestsellers throughout the Romantic and Victorian eras, competing with novels for sales.[58] And, arguably, Victorian novelists engaged more directly with science than their poetic counterparts.[59] George Eliot's novels, for instance, reference natural history, optics, positivism, and "a kind of abstract sociological descriptiveness derived from science."[60] However, "novels by Eliot, [Wilkie] Collins, Charlotte Brontë, Elizabeth Gaskell, and others are full of frustrated intellectual women," acknowledging their exclusion from the specialized ranks of science and thus, unlike some earlier writers' works, refraining from encouraging women's serious scientific participation.[61]

Throughout the nineteenth century, women continued to write religious and educational texts as well as travel narratives on nature, including Margaret Gatty's religious *Parables from Nature* (1855), Marianne North's reflections on her travels as a botanical artist in *Recollections of a Happy Life,* Arabella Buckley's Darwinian account of ants in *Life and Her Children* (1881), and Mary Kingsley's travel narratives that detail new species of fish in Africa.[62] Jane Marcet's expository books on, for instance, botany and chemistry summarized and popularized the work of Humphry Davy, who himself abandoned poetry for science in his youth; and the Scottish Mary Somerville wrote on scientific subjects including astronomy, geography, and microscopy. The respected fossil collector, dealer, and paleontologist Mary Anning was ineligible as a woman to join the Geological Society of London, but discovered the first correctly identified ichthyosaur skeleton, among other important finds. Later in the century, Eleanor Anne Ormerod gained global fame for her annual *Report on Injurious Insects* (1877–1900) and became the first woman Fellow of the Royal Meteorological Society in 1878. Yet, between the 1830s and 1880s serious or didactic contributions of natural history theories and observations seemed for poets an unpopular, impractical, and unprofitable detraction from "high" literature and their professional identities and for novelists a frustrating reminder of science's professional inaccessibility for women. In the 1880s a resurgence of women poets merging science and literature occurs, with particular attention to scientific nomenclature and Darwinian (Charles now, not Erasmus) evolution, in the verses of Constance Naden, Mathilde Blind, and May Kendall. However, five decades intervene between this later movement and the unique window of opportunity for integrating literature and natural history seized by the women explored here.

Indeed, as this study has shown, even in the early decades of the nineteenth century, literary naturalism's appeal began to wane for writers of both sexes aspiring to high poetry, partly because scientific accuracies soon appeared devoid of both "taste" and originality. As early as 1795 Anna Barbauld, in a work of literary criticism, builds on her previous doubts about the extent to which poets should versify erudite, scientific details, describing didactic poetry "as a species of inferior merit compared with those which are more peculiarly the work of the imagination."[63] With dubious praise, she exemplifies Erasmus Darwin's *Botanic Garden* as among the best of these didactic, scientific works, tepidly commending his verse depiction of Richard Arkwright's cotton-carding engine, "a piece of mechanism as complete in its kind as that which he describes."[64] She complains that in recent years "hardly any branch of knowledge has been so abstruse, or so barren of delight as not to have afforded a subject to the Didactic Poet. Even the loathsomeness of disease and the dry maxims of medical knowledge have been decorated with the charms of poetry."[65] Barbauld thus pronounces the exhaustion of scientific poetry's potential novelty and urges its abandonment for poetry of the imagination if no further didactic subject can be made "in itself attractive to the man of taste."[66]

Barbauld was not alone in her complaints, and here anticipates Charlotte Smith's late doubts about sustaining scientific literature's novelty. As I have discussed, even the writers and naturalists most earnestly invested in uniting natural history and literature often kept a sense of humor about their subject. However, especially by the early decades of the nineteenth century, in public satires of such scientific literature the laughter sounds more contemptuous than good-natured.[67] For instance, in 1812, the year following publication of Anna Seward's personal correspondence, where she disapproves of Robert Southey's stylistically experimental *Thalaba* as unclassifiable within her poetic taxonomy, Southey published his own literary taxonomy, a "Classification of Novels," in his *Omniana; or Horae Otiosiores* (1812), delineating,

> Novels may be arranged according to the botanical system of Linnaeus. Monandria Monogynia is the usual class, most novels having one hero and one heroine. Sir Charles Grandison belongs to the Monandria Digynia. Those in which the families of the two lovers are at variance may be called Dioecious. The Cryptogamia are very numerous, so are the Polygamia.—Where the lady is in doubt which of her lovers to chuse, the tale is to be classed under the Icosandria. Where the party hesitates between love and duty, or avarice and ambition, Didynamia. Many are poisonous, few of any use, and far the greater number are annuals.[68]

Southey's ridicule of novels extends to the Linnaean botanical system itself, which by this time was in growing disfavor among naturalists who viewed its artificial classifications according to plants' sexual characters as inadequate to achieve a "natural" order.[69] Moreover, Southey presents interrelations between natural history and literature or criticism more generally as absurd, voicing an increasingly widespread critical perspective that would relegate to oblivion in subsequent decades these women's intertextual pursuits of literary originality through science. When, at the end of the nineteenth century, Huxley credits Tennyson as the "only poet since the time of Lucretius, who has taken the trouble to understand the work and tendency of the men of science," he articulates the pervasive discounting and erasure of this Romantic-era scientific literature that only recently has begun to be rediscovered and revaluated.[70]

As I have shown, changing concepts of originality explain why this movement of scientific literature arose in the latter half of the eighteenth century, why these women wrote as they did, and why this literary movement failed or altered around the 1820s and '30s. This collective, collaborative mode influenced male authors of high Romanticism and, particularly as critics more strongly championed autonomous originality, the second generation of these scientific women writers, in turn, shifted toward a style that is more emotional, imaginative, and domestic, and less overtly intertextual or scientific. In the latter half of the eighteenth century and in the early nineteenth, women's literary naturalism enabled them to challenge the assertions of male naturalists and suggest women's supremacy in both poetry and science, as well as explore possibilities for literary originality. Employing natural history taxonomies as models for literary classification, they conceptualized a different literary canon while creating or discovering new literary forms and inviting readers to help solve nature's mysteries. In an era fraught with revolution and governmental anxieties, these women voiced support for various political views and social orders through scientific interpretations of nature. Evincing scientific authority, they simultaneously could envision alternative gender and social circumstances and critique masculine literary and scientific perspectives that neglected domestic, moral, or social obligations.

Although, especially after the 1820s, many factors converged to change this literary capacity, the women writers of my study employed natural history as a powerful source of cultural authority, helping to shape the Romantic era and its ideas of authorship and originality, setting the stage for succeeding generations. In so doing, they provoked a range of enthusiastic and negative responses from one another as well as from other authors, naturalists, and literary critics. Their works provide insight into the overlaps between eighteenth- and nineteenth-century literary and scientific networks and traditions during the crucial decades in which these convergences also deteriorated. Women writers' scientific literature of this period, their various textual collaborations, allow us more fully to illuminate how they influenced and defined not only their era's evolving quest for literary originality but also science's and literature's specializations into their particular modern disciplines and the creation of the current literary canon in which they must be given a more prominent place.

NOTES

✛

Introduction

1. On the phrase, "second scientific revolution," see, for instance. Holmes, *Age of Wonder,* xvi. While the "second scientific revolution" traditionally referred to a scientific movement around 1800, especially associated with chemistry, more recently scholars have expanded the concept and its temporal frame to include advancements in other developing scientific disciplines of the time.

2. Macfarlane, *Original Copy.*

3. Of course, these aren't the only scholars to whose writing on the history of science this project responds, but for a sampling of these important critics' works on natural history, see Bewell, "Jacobin Plants," 132–39; Schiebinger, *Nature's Body;* Shteir, *Cultivating Women;* Bewell, *Romanticism and Colonial Disease;* Heringman, *Romantic Rocks;* Schiebinger, *Plants and Empire;* George, *Botany, Sexuality;* O'Connor, *Earth on Show.* See also, more recently, Kelley, *Clandestine Marriage* and Ruston, *Creating Romanticism.*

4. Russell and Tuite, eds. *Romantic Sociability,* 19.

5. See, for example, George, *Botany, Sexuality,* 177. My conclusion provides greater explanation of this claim.

6. Boyle, "Astell and Cartesian 'Scientia'," illustrates that although the notion of separate spheres of activity for men and women "reached its zenith in the mid-nineteenth century," the concept can be found in the writings of Aristotle and was certainly in place throughout the eighteenth century (109).

7. Wolfson, *Borderlines,* 15.

8. See, for example, Mellor, *Mothers of the Nation;* Ahern, *Affect and Abolition in the Anglo-Atlantic.*

9. Johnson, *Adventurer,* 145.

10. Goldsmith, *Inquiry into the Present State,* 3, 195–96.

11. In addition to claiming the best subjects for poetry, according to eighteenth-century critics, these early authors enjoyed the ironic advantage of an era free of literary criticism, a practice portrayed as originating with Aristotle and as constraining verse. In his *Essay on Original Genius* (1767), Duff distinguishes Shakespeare as the only modern author whose originality overcomes these limitations to compare with that of the ancient poets (Duff, *Essay on Original Genius,* 287).

12. Bate, *Burden of the Past.*

13. Addison, *Spectator.*

14. Steele, *Guardian*.

15. Warton, *Adventurer*, 374.

16. Terry, *Plagiarism Allegation*, 6.

17. See, for example, Mazzeo, *Plagiarism and Literary Property*; Terry, *Plagiarism Allegation*.

18. Macfarlane, *Original Copy*, 41.

19. Johnson, *Adventurer*, 150; Stockdale, *Inquiry into the Nature*, 86.

20. Edward Young, *Conjectures*, 9, 27, 42.

21. Stockdale, *Inquiry into the Nature*, 60.

22. Edward Young, *Conjectures*; Goldsmith, *Inquiry into the Present State*; Stockdale, *Inquiry into the Nature*; Percival, *Moral and Literary Dissertations*; John Aikin, *Essay on the Application*.

23. See Pearson, *Women's Reading*, 67.

24. See, for example, Shteir, "Green-Stocking or Blue?," 3–14.

25. Haywood, *Female Spectator*; Lennox, *Lady's Museum*. On Lennox, see Italia, *Rise of Literary Journalism*, 203–4.

26. Haywood, *Female Spectator*, 3.154, 296; 4.48; Italia, *Rise of Literary Journalism*, 203.

27. Haywood, *Female Spectator*, 3.155, 147, 294.

28. See Girten, "Unsexed Souls," 55–74.

29. Haywood, *Female Spectator*, 3.156, 158.

30. Roger, *Buffon*, 311–12.

31. Also, as Heringman ("Natural History in the Romantic Period," 145), among others, demarcates, "scientists" began being designated by this term in 1833.

32. Crawford, *Poetry, Enclosure, and the Vernacular Landscape*, 92.

33. Feingold, *Nature and Society*, 1; Crawford, *Poetry, Enclosure, and the Vernacular Landscape*, 92; Siskin, *Historicity of Romantic Discourse*, 12; and Goodman, *Georgic Modernity*, 10.

34. John Aikin, *Essay on the Application*, 41.

35. Ibid., 59.

36. Goldsmith, *Inquiry into the Present State*, 150.

37. Heron, *Letters of Literature*, 93, 94.

38. Stockdale, *Inquiry into the Nature*, 102–3.

39. Ibid., 104, 161.

40. Goodman, *Georgic Modernity*, 30.

41. Qtd in Chalker, *English Georgic*, 59.

42. Trapp, "Of Didactic or Preceptive Poetry," 188.

43. Duff, *Essay on Original Genius*, 274.

44. Ibid., 275.

45. Ibid., 277.

46. Clymer, introduction, 3.

47. Duff, *Essay on Original Genius*, 275.

48. Numerous books have been written on individual male Romantic authors' interactions with science; however, the era's only female writer thus far to be given this level of scholarly attention for her scientific interests is of course

Mary Shelley. For some of these studies on individual male Romantic authors' engagements with science, see, for example, Thomas and Ober, *Mind For Ever Voyaging*; Wyatt, *Wordsworth and the Geologists*; Roe, *Samuel Taylor Coleridge*; Pamela Edwards, *Statesman's Science*; Levere, *Poetry Realized in Nature*; Peterfreund, *William Blake in a Newtonian World*; Ruston, *Shelley and Vitality*; Goellnicht, *Poet-Physician*; Fara, *Erasmus Darwin*; Berns, *Science, Politics, and Friendship*.

49. Pascoe, "Unsex'd Females," 211.

50. Schiebinger, *Mind Has No Sex?*, 245; Hayden, "Introduction," 5; Shteir, *Cultivating Women*; George, *Botany, Sexuality*; Pascoe, "Female Botanists," 193–209.

51. To name a few critics who have written in this vein, see Macfarlane, *Original Copy*; Mazzeo, *Plagiarism and Literary Property*; Leader, *Revision and Romantic Authorship*; Stillinger, *Multiple Authorship*.

52. Wordsworth, *Prose Works*, 3:26; "Who shall point as with a wand, and say/'This portion of the river of my mind/Came from yon fountain'?" (Wordsworth, *The Prelude*, 84).

53. Percy Shelley, *Shelley's Prose*, 293, 328. See also Macfarlane, *Original Copy*, 31.

54. Hazlitt, *Lectures on the English Poets,* 322, 323.

55. Ibid., 323.

56. Catherine Ross, "Twin Labourers," 23–52. Shteir writes of the branches of natural history that, while botany remained "the most suitable for 'ladies'; zoology was the least suitable, with mineralogy and conchology in between" ("Green-Stocking or Blue?," 7). Conchology received notable attention from women writers such as Sarah Hoare, and from famous female personalities including the Duchess of Portland.

57. Fulford, Lee, and Kitson, *Literature, Science, and Exploration,* 4.

58. I discuss Wordsworth's preface in more detail in chapter 1, and Byron's interactions with women and natural history receive treatment in chapters 5 and 6; for Clare's resistance to Linnaean thought, see Kelley, "Romantic Exemplarity," 223–54.

59. See, for example, Pascoe, "Female Botanists"; Mahood, *Poet as Botanist,* 20.

60. For instance, Browne, "Botany for Gentlemen," 593–621; Uglow, "But What about the Women?," 163–78.

61. As with most popular literary movements, this melding of natural history and literature became a site for satire almost as immediately as it began. Drawing on Linnaeus's "sexual system," Vincent Miller's *The Man Plant: Or, Scheme for Increasing and Improving the British Breed* (1752) mocks Linnaean analogies between zoological and botanical kingdoms, likening a gardener's daughter, Sally, to a plant, seduced and productive of a human egg to be planted in the earth for eight months and taxonomized in the class *Dioecia*, order *Monandria*, and genus *Homo*. In this vein, James Perry's satirical verses *Mimosa* (1779), on the sensitive plant or *Mimosa pudica* served as a thinly veiled reference to the penis. John Wilcot, under the pseudonym Peter Pindar, published several literary caricatures of the naturalist Joseph Banks, including "Sir Joseph Banks and the Boiled Fleas" (1790), in which Banks attempts to prove that fleas are really lobsters, thus emphasizing the frivolity of his scientific inquiries. In 1798 George Canning's *The Loves of the Triangles* parodied Darwin's *The Loves of the Plants*, greatly damaging the

latter's ability to be taken seriously. Numerous additional parodies of contemporary scientific literature could be listed here and, to state one more, Thomas Love Peacock's *Melincourt* (1817) offers tongue-in-cheek use of learned footnotes in his story of an orangutan, Sir Oran Haut-Ton, put forward as a candidate for election as Member of Parliament. For more on the era's scientific satires, see, for example, Lawlor, "Poetry and Science," 47–48.

1. To Teach and to Please

1. Lamb, *Letters of Charles and Mary Anne Lamb*, 2:82.

2. Mandell, *Misogynous Economies*, 129.

3. Barbauld, *Anna Letitia Barbauld: Selected Poetry and Prose*, 480. Henceforward, unless otherwise documented, all references to Barbauld's poetry and prose will be cited from this work and appear by line or page number directly in this chapter's text.

4. McCarthy, *Anna Letitia Barbauld*, 49. The character of Charles recurs in Barbauld's writings for children, most prominently in *Lessons for Children*. The character Charles is the namesake of a real-life counterpart, her nephew and adopted son—the biological son of her brother, John Aikin.

5. Aikin and Barbauld, *Evenings at Home*, 2:132, 134.

6. Ibid., 2:136.

7. Ibid., 2:134.

8. Ibid., 2:135–36.

9. John Aikin, *Essay on the Application*, 25.

10. Plank, "John Aikin on Science and Poetry," 168.

11. Pennant, *British Zoology*, xiii.

12. John Aikin, *Essay on the Application*, 48.

13. Ibid., 122–23.

14. Butterflies were traditionally associated with souls in Greek mythology; and, in ancient Greek, the word for butterfly is *psyche*, which translates to "soul."

15. John Aikin, *Essay on the Plan and Character*, xv–xvi.

16. *Monthly Review*, 48 (1773): 58.

17. Lucy Aikin, *Works of Anna Letitia Barbauld*, 2:15.

18. Here, it is interesting to wonder if the verses Barbauld had in mind might have been James Grainger's *The Sugar-Cane* (1764), treated in the next chapter. As Gilmore relates, a contemporary writer, "Griffith Hughes, was even criticized for using common names in a prose work on the natural history of a Caribbean island, names which were criticized for being not only English rather than Latin, but also 'the very worst English of West India planters'," so that "when we find Grainger introducing yams, okras, and bonavist into his georgic, there is more than 'novelty' to it, there is artistic boldness which should not be underestimated" (*The Poetics of Empire*, 32).

19. John Aikin, *Essay on the Application*, 147.

20. Lucy Aikin, *Works of Anna Letitia Barbauld*, 2:16.

21. Barbauld, *Anna Letitia Barbauld*, 46n1.

22. Barbauld, *Pleasures of Imagination*, 2–3.

23. Ibid., 1.

24. John Aikin, *Calendar of Nature,* iii.

25. This reference to creation is in Linnaeus's *Amoenitates.* Thanks to the Linnean Society of London for access to this correspondence. See also McCarthy, *Anna Letitia Barbauld,* 43.

26. Lucy Aikin, *Works of Anna Letitia Barbauld,* 1:xxvii.

27. Barbauld, *Anna Letitia Barbauld,* 43n.4.

28. On several occasions Barbauld expressed admiration for William Paley's treatise *Natural Theology* (1802), which borrowed extensively from Ray's earlier text. She lauds Paley, perhaps most memorably, in her poem "Eighteen Hundred and Eleven" (1812), where she imagines future American children benefiting from Britain's then-ancient intellectual tradition so that "Thy [Britain's] Lockes, thy Paleys shall instruct their youth,/Thy leading star direct their search for truth" (ll. 87–90). Nevertheless, Barbauld incorporated natural theology into her works long before Paley's book hit the press.

29. Ray, *Wisdom of God,* 111.

30. Ibid.

31. Ibid., 112.

32. Writing to Pulteney, Aikin praised the progress of a particular student in botany demonstrating this discipline's importance to the school's curriculum (letter 9, Warrington, June 30, 1760, Linnean Society of London). Barbauld recommended this method of learning in her letters "On Female Studies" (Barbauld, *Anna Letitia Barbauld,* 482).

33. Barbauld, *Poems of Anna Letitia Barbauld,* 227. Barbauld had to convince her father to let her study these classical languages; see Lucy Aikin, *Works of Anna Letitia Barbauld,* 1:vii.

34. Further demonstrating Barbauld's sense of humor and appreciation of this science's entertainment value, she inserted elaborate natural-historical comparisons and references into her letters that show her wry wit by, for instance, likening her lack of intellectual productivity to the state of an animal in hibernation: "The only difference being, that *I* have all the while continued the habit of eating and drinking, which, to their advantage, *they* can dispense with" (Lucy Aikin, *Works of Anna Letitia Barbauld,* 2:97); or conjecturing about the transmigrating spirit of her turkey dinner: "I hope it is animating some other vehicle, and rising by degrees in the scale of existence, till perhaps it may come at length (who knows) to eat turkey itself" (ibid., 2:98); or pardoning infrequent correspondence with a friend by admitting that her motivation to write "is rather like the aloe, that after having been barren season after season shows signs of life all on a sudden, and pushes out when you least expect it" (ibid., 2:68)

35. Eagerly keeping abreast of the scientific advancements made in her age, Barbauld sometimes attended lectures at the Royal Institution, as on one occasion when she "was much pleased to see a fashionable and very attentive audience, about one third ladies, assembled for the purposes of science and improvement" (Lucy Aikin, *Works of Anna Letitia Barbauld,* 2:67).

36. Schiebinger, "Gender and Natural History," 163–4.

37. For more analysis of Linnaeus's "sexual system" see, for example, Browne, "Botany for Gentlemen"; Shteir, *Cultivating Women.*

38. Wollstonecraft, *Critical Edition*, 122. Kelley also analyzes this botanical contention between Wollstonecraft and Barbauld in *Clandestine Marriage*, 99–103.

39. Page and Smith, *Women, Literature, and the Domesticated Landscape*, 92; Shteir, *Cultivating Women*, 36.

40. I refer to Keats's poetic "axioms" as expressed in his letter to John Taylor, February 27, 1818 (Keats, *Letters*, 69–70).

41. This is a quote from Barbauld's "The Rights of Woman" (l. 14). As Curran notes, Barbauld strives for "liberation through, not from, femininity" ("The I Altered," 197).

42. McCarthy, *Anna Letitia Barbauld*, 119–20.

43. *Monthly Review* 48 (1773): 133. Also, see McCarthy's book article on this subject, "We Hoped the Woman was Going to Appear," 113–37.

44. See Mandell, *Misogynous Economies*, 151. Barbauld felt comfortable in critical evaluation of scientific as well as literary texts by her contemporaries. For instance, she critiqued P. Darling's *Natural History of Quadrupeds, Paternal Love* for the *Monthly Review* (McCarthy, *Anna Letitia Barbauld*, 516).

45. See, for example, George, *Botany, Sexuality*, 111.

46. Darwin, *Botanic Garden* (1789), ix, vi.

47. Seward, *Letters*, 6:157.

48. Dyce, *Recollections*, 182.

49. Pennant, *British Zoology*, xiv. Lucy Aikin, *Works of Anna Letitia Barbauld*, 2:16.

50. This interlude's dialogue begins with the Bookseller's complaint that "Your verses, Mr. Botanist, consist of *pure description*, I hope there is *sense* in the notes" (Darwin, *Botanic Garden* [1789], 40). Darwin's italicization of "pure description" and "sense" echoes an assertion in John Aikin's essay of literary criticism on Thomson's *The Seasons*, published the previous year. There, Aikin demonstrates concern for legitimizing descriptive poetry, noting that prior to Thomson's poem it was assumed that, in poetical practice, "*pure description* was opposed to *sense*; and binding together the wild flowers which grew obvious to common sight and touch, was deemed a trifling and unprofitable amusement" (*Essay on the Plan and Character*, vii). For Aikin, Thomson succeeded in legitimizing descriptive poetry by fashioning a "progressive series of descriptions" that are not arbitrary, but contribute to "a general plan" (ix). Thus when Darwin's Poet claims, "I am only a flower-painter, or occasionally attempt a landskip," he lacks Aikin's anxiety about proving that descriptive verse does not preclude "sense"; instead, Darwin embraces their separation.

51. Darwin, *Botanic Garden* (1789), 41, 43.

52. Mahood, *Poet as Botanist*, 56.

53. My argument in this paragraph is indebted to Roger, *Buffon*, 309–16. This argument could, in reality, be dated back to Aristotle, but I am briefly charting its manifestations in the long eighteenth century.

54. Ray, and later Buffon, gave up the search for an "essential characteristic" and instead based the criterion for a species on the ability to "perpetuate and conserve similarity of the species by means of copulation," Roger, *Buffon*, 313–14.

55. Curran, "The I Altered," 189.

56. Barbauld, "Prefatory Essay," xiv, xxv.

57. Ibid., iii.

58. Ibid., iii–iv.

59. Ibid., iv, v.

60. Ibid., vi.

61. Ibid., iv, vi, vii.

62. Ibid., vi.

63. A skill noted, for instance, by writers such as Eliza Haywood and Charlotte Lennox (see introduction).

64. Qtd in Barrell, *Political Theory of Painting,* 66.

65. For two sensitive readings of this poem, see Doody, "Sensuousness in the Poetry," 24–25; Pascoe, "Unsex'd Females," 222–23.

66. Pascoe, "Female Botanists," 193.

67. Hartman, *Wordsworth's Poetry,* 5.

68. Wordsworth, *William Wordsworth,* 606–7.

69. Catherine Ross, "Twin Labourers," 23–52; Bewell, "Erasmus Darwin's Cosmopolitan Nature," 20.

2. Hybrid Britons

1. Johnson, *Idler.*

2. For more on the georgic as a hybrid form, see Morris, *Scotland and the Caribbean,* 71–2.

3. Riddell, *Voyages.* Henceforward, page numbers from this work will be cited parenthetically within this chapter's text.

4. See especially Bewell, *Romanticism and Colonial Disease,* and "Jefferson's Thermometer"; Grove, *Green Imperialism*; Parrish, *American Curiosity*; Wheeler, *Complexion of Race*; and Wilson, *Island Race.*

5. For Riddell's biographical information, see Macnaghten, *Burns' Mrs. Riddell*; and Gladstone, *Maria Riddell.*

6. Riddell, *Metrical Miscellany,* 69–71.

7. While Albion often poetically refers to England or Great Britain, the term historically signifies Scotland.

8. McIntyre, *Dirt and Deity,* 354.

9. Ibid., 349.

10. Kerr, *Memoirs,* 2:362–63.

11. More, *Strictures,* 2:12.

12. Kerr, *Memoirs,* 2:365.

13. *Analytical Review* 16 (1793): 375; *Gentleman's Magazine* 63 (1793): 1.

14. Stewart, *Elements of the Natural History,* 1:13.

15. Katherine Turner, *British Travel Writers,* 181.

16. Schaw, *Journal of a Lady of Quality.* Turner remarks that although Schaw's travel narrative was not published within her lifetime, her manuscripts suggest that the work was intended for a wide readership and that Schaw may have been "considering publication," Katherine Turner, *British Travel Writers,* 134.

17. Kerr, *Memoirs,* 2:363.

18. O'Shaughnessy, *Empire Divided,* 4, 208.

19. Qtd in O'Shaughnessy, *Empire Divided,* 247.

20. For the various reasons scholars offer in determining the cause for the georgic's disappearance, see the introduction.

21. Grainger, *Sugar-Cane*.

22. Ibid., v.

23. Ibid., vii.

24. Grainger became related by marriage to James Verchild, the president of St. Kitts and later governor of the Leewards; Gilmore, *Poetics of Empire*, 13.

25. See Gilmore's note to Grainger's lines, 1:329–30, in *Poetics of Empire*, 228–29.

26. Crawford, *Poetry, Enclosure, and the Vernacular Landscape*, 92.

27. Grainger, *Sugar-Cane*, 4:676.

28. Ibid., 4:235–36.

29. Ibid., 3:166–79, 4:34–102, 4:151. Grainger describes this "toil" as generally welcome to "the Negroe-train" who "pant to wield the bill" with "willing ardour," for "cheerful toil is light" (3:96, 98, 100, 101). He thus participates, and fails, in the generic convention outlined by Beth Tobin, in which georgic poets "valorize intellectual labor, . . . [and] employed the trope of bounty to undo the power of direct producers—the peasants, slaves, and natives who grew plants and raised animals," portraying that physical labor as "light" (Tobin, *Colonizing Nature*, 32).

30. Frohock, *Heroes of Empire*, 179.

31. Thomson, "Rule, Brittania" (1740) l. 6, in Thomson, *Complete Poetical Works*.

32. Crawford, *Poetry, Enclosure, and the Vernacular Landscape*, 94.

33. Sambrook, *James Thomson*, 53.

34. Tara Wallace, *Imperial Characters*, 53. According to Kaul, Thomson establishes this unifying perspective as "the best way of charting an uncertain, but nevertheless very attractive, imperial future" (*Poems of Nation*, 144).

35. Thomson, *The Seasons*, "Summer," ll. 192–96, in Thomson, *Complete Poetical Works*.

36. Thomson, *The Seasons*, "Spring," ll. 221–23, in Thomson, *Complete Poetical Works*.

37. Riddell does follow Linnaean taxonomy in her classifications of insects, as these species are not included in Pennant's system; see Pennant, *British Zoology*; Pennant, *History of Quadrupeds*.

38. Pennant, *History of Quadrupeds*, 1:iii.

39. Ibid., 1:iii–iv.

40. Pennant, *British Zoology*, 1:xix.

41. Gladstone, *Maria Riddell*, 29.

42. Pennant, *Tour of Scotland*; *Tour in Wales*; *Journey from Chester to London*; *Of London*; *Literary Life*.

43. Gottlieb, *Feeling British*, 18.

44. Pennant, *Literary Life*, 13.

45. Ibid., 135.

46. Pennant, *British Zoology*, 1:xiii. Pennant also discusses his adoption of Ray's system in the preface of *History of Quadrupeds*, and it is in this work that Pennant more specifically sets up the rivalry between himself, Linnaeus, and Buffon.

47. Pennant, *British Zoology,* 1:iv, xi.

48. Ibid., 1:vi. Pennant appeals to natural theology in the respective works of Robert Boyle, John Ray, and William Derham. He emphasizes Protestantism as a notably British value, for instance, in his approving quotation of "A judicious Foreigner" who remarked "that an *Englishman* is excusable should he be ignorant of papal history where it does not relate to Great Britain" (iii). Pennant urges concern with that which he believes does directly relate to Britain, particularly the economic potential of its natural history.

49. Great Britain "was an invention forged above all by war. Time and time again, war with France brought Britons whether they hailed from Wales or Scotland or England, into confrontation with an obviously hostile Other and encouraged them to define themselves as Protestants struggling for survival against the world's foremost Catholic power" (Colley, *Britons,* 5).

50. Pennant, *British Zoology,* 1:xi.

51. Edwards attempts a more overt effort at conciliation between Britain and its West Indian colonies in his text *History, Civil and Commercial,* 2:477.

52. The Catholic Irish may seem to be an exception to British Protestant unification forged through opposition to the Catholic "other," but in the West Indies, "with the scarcity of Roman Catholic priests, many of the Irish gradually conformed to the Church of England, or (in the Leeward Islands) moved to French territory and became identified with the French" (Dayfoot, *Shaping of the West Indian Church,* 86).

53. For the contemporary controversy surrounding translation of this line (l. 149) from Virgil's *Georgicorum, Liber Secundis,* see Martyn, *Georgicks of Virgil,* 171.

54. Cousins, "Marvell's Devout Mythology," 216.

55. Ibid., 217.

56. Low, *Georgic Revolution,* 255.

57. Armitage, *Ideological Origins,* 97–98.

58. Gregory, *Beneficent Usurpers,* 18. British ships regularly stopped at Madeira to take on wine for their voyages, and Madeira wine was especially popular in the West Indies because it improved in high temperatures.

59. Many scholars have remarked on the ubiquitous despondent nuns in women's travel writing of this time; see, for instance, Bending, *Green Retreats,* 71.

60. Born, *John Physiophilus's Specimen of the Natural History,* 5.

61. Ibid., 14–15.

62. Brown, *Reaper's Garden.*

63. Gregory, *Beneficent Usurpers,* 78.

64. Kerr, *Memoirs,* 2:391. While this quote is taken from a letter Riddell wrote a few years after the publication of *Voyages* and pertains to events transpiring in Britain itself, her reactions to the Catholic funeral and to the Catholic religion more generally on Madeira suggest the quote's applicability here. Also, when she draws attention to the skin color of Madeiran Portuguese as "extremely dark complexioned," she seems to echo Long's earlier indictments of a degenerative "link between complexion and moral probity" (*History of Jamaica,* 14; see also Wheeler, *Complexion of Race,* 141). While Riddell may advocate this correlation

between race and morality in the Portuguese, she does not apply it to inhabitants of the West Indies (slave or free).

65. Schaw, *Journal of a Lady of Quality*, 74.

66. See, for example, Charlotte Smith's attention to such species in her poem "Flora" (in Smith, *Poems*).

67. For instance, during the period of St. Kitts's history between 1627 and 1713, the island was shared between the English and the French, an uncomfortable and unstable compromise of insular hybridity that provoked each nation to invade the other half of the island in alternating successions before St. Kitts finally came into Britain's definitive possession.

68. Hollocher, "Island Hopping in *Drosophilia*," 131.

69. Losos, "Ecological and Evolutionary Determinants," 210–24; Grant, introduction, 8–9.

70. Smellie uses these terms interchangeably in the section titled "Of the Sexes of Plants" in his *Philosophy of Natural History*. Also, in *Sugar-Cane*, Grainger frequently mentions mules as integral to the process of producing sugar in the British West Indies, but Riddell neglects such reference to the sugar industry. For further discussion of West Indian and colonial hybridity, see Casid, *Sowing Empire*, 1–44.

71. Buffon, *Natural History*, 7:445.

72. Ibid., 8:15–17.

73. Ibid., 8:34.

74. Bryan Edwards, *History, Civil and Commercial*, 1:viii–ix.

75. Buffon, *Natural History*, 7:448. Buffon noted "improvement" in "four or five species of polecats" in the Americas, and wrote that "the roebucks and the fallow-deer, as well as the mouffettes, are more numerous and likewise stronger in the New than in the Old Continent." See also Bewell's *Romanticism and Colonial Disease* and Grove's *Green Imperialism* for studies of eighteenth- and early nineteenth-century thought regarding the potential of land cultivation to "improve" both natural environments and climate.

76. Buffon, *Natural History*, 7:399.

77. Sheep play a key role in Buffon's chapter "Of the Degeneration of Animals"; interestingly, Riddell's description of the alteration undergone by European sheep brought to the West Indies is similar to Buffon's description of the species when "restored" to its original form; see *Natural History*, 7:398–400.

78. Long, *History of Jamaica*, 2:261; Bryan Edwards, *History, Civil and Commercial*, 2:10.

79. Qtd in Roger, *Buffon*, 263.

80. Long, *History of Jamaica*, 1:441.

81. Crèvecoeur, *Letters from an American Farmer*, 175–76.

82. Long, *History of Jamaica*, 1:437.

83. Wheeler, *Complexion of Race*, 210.

84. Long, *History of Jamaica*, 2:327; see also Robert Young, *Colonial Desire*, 8.

85. Kerr, *Memoirs*, 2:375–76. It is also notable that, in this letter, Riddell more candidly acknowledges the devastations of West Indian disease, and specifically of yellow fever, than she does in *Voyages*. In 1802 Riddell's husband died on Antigua, presumably of disease; Gladstone, *Maria Riddell*, 38.

86. In "Scarred, Suffering Bodies," Andrews agrees that Riddell is influenced by her family's ownership of plantations in the West Indies, but sees no "agenda" in her elision of slavery, simply attributing it to her lack of "moral capital" (187, 189).

87. Bryan Edwards, *History, Civil and Commercial,* 2:7–8. 3.

3. The Evolution of the Plagiarist

1. Seward, *Letters,* 2:287, 1:163, 2:162. Henceforward, letters from this collection will be cited in the text of this chapter by volume and page number. Seward carefully prepared these letters to be published upon her death. Although there remains far less scholarship on Seward than on Smith, there recently has been a surge in Seward criticism, including two important critical biographies: Barnard, *Anna Seward: A Constructed Life*; and Kairoff, *Anna Seward and the End of the Eighteenth Century.*

2. Pinch, *Strange Fits,* 61. For a survey of recent critical attention to Smith's poetical borrowings, see Backscheider, *Eighteenth-Century Women Poets,* 335–36.

3. See, for instance, Robinson, "Reviving the Sonnet," 112–13.

4. Seward thus participates in what Pascoe has "recognized as a literary movement . . . of British women's writing merging poetry and science" in her essay "Female Botanists," 193.

5. This is a phrase from Seward's sonnet "To Mr. Henry Cary, on the Publication of his Sonnets," first published in 1788. Robinson makes extensive use of this phrase and explains that "the sonnet is a form that women writers deliberately claimed in order to legitimize themselves as poets" ("Reviving the Sonnet," 99).

6. Fletcher, *Charlotte Smith,* 327.

7. *Critical Review,* 3rd ser., 2 (1804): 198.

8. For studies of controversies surrounding women's education in botany, the most popular division of natural history, see Shteir, *Cultivating Women;* and Bewell, "Jacobin Plants," 132–39. See also George, *Botany, Sexuality.*

9. Seward, *Memoirs of the Life of Dr. Darwin,* 217.

10. Ibid., 131.

11. Poems in which Seward incorporates elements of natural history or natural philosophy include *Elegy on Captain Cook,* "Bermuda," "The Terrestrial Year," and "Colebrook Dale," among others. For analysis of Seward's poetic engagements with astronomy, see Barnard, "Anna Seward's 'Terrestrial Year'," 3–17.

12. Seward, *Memoirs of the Life of Dr. Darwin,* 362, 336.

13. Exemplifying her faith in her critical capacities, Seward, in a letter quoted in *Poetical Works of Anna Seward, Edited by Walter Scott,* commented: "Many excel me in the power of writing verse; perhaps scarcely one in the vivid and strong sensibility of its excellence, or in the ability to estimate its claims" (1:xiii).

14. Bewell, "On the Banks of the South Sea," 175. See also Schiebinger, *Nature's Body,* 9.

15. Withering, "Introduction to the Study of Botany," 1:6. Withering and Darwin were both members of the Lunar Society of Birmingham. In *Lunar Men,* Uglow relates two instances in which Darwin appropriated Withering's work. In

the first instance, Withering taught Darwin a cure he discovered through treatment with digitalis, and Darwin then published this information in the name of his deceased son. After this occurrence, Withering hardly spoke to Darwin again; however, this did not stop Darwin from "stealing [Withering's] pronunciation scheme" of botany a few years later for inclusion in Darwin's translation of Linnaeus. Withering was furious, but "Darwin simply shrugged off the dispute as if it was not worth bothering with" (Uglow, *Lunar Men*, 279, 381–82).

16. Trevor Ross, "Two Ways of Looking at a Canon," 91. Thanks to Ted Underwood for alerting me to this source. While Ross places the hierarchical canon's inception more generally in the eighteenth century, his chief examples (both here and in his text cited below) are from the century's second half, particularly in the efforts of Joseph and Thomas Warton.

17. Trevor Ross, *Making of the English Literary Canon*, 253, 255.

18. Sloan, "Gaze of Natural History," 134.

19. Seward, *Poetical Works*, 1:lxxxiii.

20. Seward, *Memoirs of the Life of Dr. Darwin*, 394–95.

21. Ibid., 180–81.

22. Ibid., 400.

23. Ibid., 132.

24. Part of what makes Darwin's borrowing of Seward's lines an instance of plagiarism is that her poem had already "passed the press in the name of their real author." Of course, as Barnard and DeLucia, for instance, have shown, Seward and numerous other poets offered suggestions about word choice, lines, and ideas to fellow poets who solicited advice about their works before publication (Barnard, "Anna Seward's Hidden Words," 417–33; DeLucia, "Local Poetry in the Midlands," 159). If the original author accepted this aid, to maintain a successful, coherent style, she must then, as Sir Walter Scott said of lines he accepted from Seward, "adopt them but not without some disfigurement and alteration, as gypsies stain stolen children to make them indistinguishable from their own" (letter to Anna Seward, May 27, 1808, ms. 3653, National Library of Scotland).

25. Mazzeo, *Plagiarism and Literary Property*, 41–42.

26. Ibid., 2.

27. Ibid., 3–4.

28. Darwin, *Botanic Garden* (1789), 132. Darwin's language of "inclosure" conjures up John Locke's discussion of property rights in the second of his *Two Treatises of Government* (1690). Locke's essay on property remained a touchstone for critical and legal disputes of plagiarism well into the nineteenth century, "explain[ing] one of the central metaphors employed by [Romantic-period] writers in bringing charges of illegitimate appropriation: the metaphor of the literary estate" (Mazzeo, *Plagiarism and Literary Property*, 11). Darwin's natural description of "single words" and "common flowers of speech" as "lawful game" pays tribute to the Lockean notion of property legitimately acquired according to the laws of labor and improvement: "He that is nourished by Acorns he pickt up under an Oak, or the Apples he gathered from the Trees in the Wood, has certainly appropriated them to himself. . . . The *labour* that was mine, removing them out of that

common state they were in, hath *fixed* my *Property* in them" (Locke, *Two Treatises of Government*, 288–89). Just as Locke's stipulation of "improvement" justified the enclosure of estates and England's imperial appropriation of foreign lands, so did successful improvement justify writers in making similar literary appropriations from their peers.

29. Darwin, *Botanic Garden* (1789), 132.

30. Edward Young, *Conjectures,* 10. Unlike Darwin, Young incorporates Lockean tenets of property only to denigrate those most applicable to literary disputes, describing "*Imitations*" as "a sort of *Manufacture* wrought up by . . . *Labour*, out of pre-existent materials not their own." Labor here becomes harmful to composition, a detriment to originality, so that according to Young's argument, Darwin's labor in the *Botanic Garden*, and even the very site of the garden, aligns his work with cultivation and artificiality that opposes the natural growth of genius.

31. Illustrating Seward's awareness of Darwin's scientific imitations, near the end of her discussion of Darwin's *Phytologia,* which hypothesizes the sentiency of plants, she clarifies, "Of this theory, however, Dr. Darwin is neither the source, nor the first who drew the scattered hints of former philosophers concerning it, into a regular system. The ingenious and excellent Dr. Percival, of Manchester, preceded him in maintaining that system from the press" (*Memoirs of the Life of Dr. Darwin*, 413).

32. King-Hele, *Erasmus Darwin,* 297. Although Linnaeus propounded the fixity of species for most of his life and was thus generally associated with this idea, Koerner discusses the doubts that later crept into his thinking: "in old age Linnaeus repented of the 162nd aphorism of *Philosophia botanica* ('the number of species is constant'), and of his famous assertion in the preface to *Systema naturae* ('We count so many species as there were in the beginning'). As he pondered the mysteries of generation, he came to feel that hybrid cross-breeding explained the earth's variety of life forms" (*Linnaeus*, 44).

33. Darwin, *Zoonomia,* 183.

34. Seward, *Memoirs of the Life of Dr. Darwin,* 87.

35. Ibid., 88.

36. Ibid., 93.

37. Sloan, "Gaze of Natural History," 121, 138. See also Roger, *Buffon,* 240–44. According to Koerner, Linnaeus believed animals possessed souls, and yet he rejected an eternal afterlife (for humans or otherwise) beyond having blood descendants (*Linnaeus*, 89). Seward contemplates the possibility of an afterlife for animals in her poem "On the Future Existence of Brutes," apparently differentiating between reason and "spirit."

38. Abrams, *Mirror and the Lamp,* 197.

39. Ibid.

40. In *Burns' Mrs. Riddell* (44), Macnaghten relates that Robert Burns adopted this line from Milton as the motto of his self-created seal. In his famous poem of 1798, *The Unsex'd Females*, Richard Polwhele portrays Ann Yearsley as a poet "who had warbled, Nature's child, / Midst twilight dews, her minstrel ditties wild" (ll. 99–100). Wordsworth also played on this analogy when he called Byron "the

Mocking Bird of our Parnassian Ornithology" (qtd. in Mazzeo, *Plagiarism and Literary Property*, 44).

41. Darwin, *Zoonomia*, 154–55; Seward, *Memoirs of the Life of Dr. Darwin*, 90.

42. Edgeworth, *Memoirs of Richard Lovell Edgeworth*, 2:266–68. Animosity subsisted between Edgeworth and Seward after his marriage to Honora Sneyd, Seward's beloved friend from youth. For more on Seward's relationship with Sneyd see, for instance, Moore, *Sister Arts*, 93–107.

43. Mazzeo, *Plagiarism and Literary Property*, 49.

44. Ibid., 53.

45. Seward desired this appreciation of merit distinct from sex in literary criticism as well as in works of literature. She reveled in the "cross-dressing" possibilities of literary criticism, sending several letters to the *Gentleman's Magazine* under the pseudonym "Benvolio" advising readers, "be it remembered that souls are of no sex, and their effusions therefore may, at pleasure, assume a masculine or a feminine application" (*Gentleman's Magazine* 57 [August 9, 1787]: 685); also quoted in Ashmun, *Singing Swan*, 142.

46. See Wood, "The Female Penseroso," 451–77.

47. Anderson, "*Beachy Head*," 123.

48. Pinch, *Strange Fits;* Anderson, "*Beachy Head*"; Backscheider, *Eighteenth-Century Women Poets*; Wolfson, "Charlotte Smith's *Emigrants*."

49. Macfarlane, *Original Copy*, 32.

50. King-Hele, *Erasmus Darwin*, 300.

51. Robinson, "Reviving the Sonnet," 100.

52. Knight, *Ordering the World*, 58–59.

53. Seward, *Memoirs of the Life of Dr. Darwin*, 178. In *Nature's Body*, Schiebinger remarks that "though Anna Seward, a close friend and well-known poet, praised *The Botanic Garden* for establishing a new poetic form by adapting scientific discoveries to heroic verse, Darwin's poetry was not new, nor was it esoteric or unusual. The eighteenth century abounded with didactic poems on raising hops, sugar cane, gardening and the like" (31). But regardless of the actual innovation of Darwin's poetry, I am interested in Seward's conviction that his poetry "forms a new class."

54. Examining the relationship between nature and poetic form, Keith writes, "For poets of the Restoration and eighteenth century, to imitate Nature is to imitate an ordering principle. Not simply the world of flora and fauna, *Nature* infinitely expands to include orders of being and knowing. Nature is a complex notion of form that defines the object of representation, the act of representation, and the subject of representation" (*Poetry and the Feminine*, 14).

55. Quoted in Clarke, "Anna Seward," 35.

56. Originality was of paramount concern to Seward in her writings. She explained, "I have always destroyed every little production of my own, if, on revising it, after the effervescence of composition had subsided, I could not find that it contained something original, either in the thoughts themselves, or in their combination" (*Letters*, 5:378).

57. Sloan, "Gaze of Natural History," 135.

58. Seward, preface to *Original Sonnets*, iii.

59. Ibid., iv.

60. For further discussion of "composite orders," see Curran, *Poetic Form,* 180–203.

61. Smith's *Elegiac Sonnets* received a great deal of critical favor, and her acknowledgments of borrowings in her prefaces and notes helped prevent accusations of plagiarism. However, as Adela Pinch has noted, even her obituary, printed by the *European Magazine,* which had long championed Smith, displays attempts to quell accusations of plagiarism, stating, for example, "Neither does her poetry always . . . display those marks of imitation that may sometimes be discovered in the works of those whose names have attained still greater celebrity" (Nov. 1806). Earlier in her career, Smith had to withdraw her translation *Manon Lescaut* following charges that the work was immoral and plagiarized. Also, readers sometimes wrote to periodicals clarifying the originations of her borrowings as, for instance, in the *Gentleman's Magazine* 56 (Sept. 1786): 757. Throughout her career, as the next chapter shows, Smith felt a constant need to defend herself against potential charges of plagiarism.

4. Plagiarism and the Poet-Naturalist

1. Smith, *Collected Letters,* 618, 646, 741.

2. Shakespeare also was known as "the swan of Avon." See Clarke, "Anna Seward," 34.

3. Smith, *Natural History of Birds,* 1:91.

4. See Wood, "Female Penseroso," 451–77.

5. Chorley, *Memorials,* 1:236.

6. Smith, *Natural History of Birds,* 1:95–96.

7. Ibid., 2:80–81.

8. Seward, *Letters,* 2:287.

9. Smith, *Natural History of Birds,* 2:94–96. I refer to Wordsworth's now well-known remark that Smith was a poet "to whom English verse is under greater obligations than are likely to be either acknowledged or remembered," quoted in the introduction of Curran's edition of Smith, *Poems,* xix.

10. Seward, *Letters,* 2:162.

11. Labbe, *Charlotte Smith,* 8.

12. Essays addressing Charlotte Smith's writings in relation to natural history include Pascoe, "Female Botanists"; Ruwe, "Charlotte Smith's Sublime," 117–32; Landry, "Green Languages?"; Anne Wallace, "Picturesque Fossils," 77–93; Tayebi, "Undermining the Eighteenth-Century Pastoral," 131–50; Cook, "Charlotte Smith," 48–67; Mellor, "Baffling Swallow," 299–309.

13. In earlier editions, Smith did include a few footnotes, citing poets and less detailed scientific information but, from the third edition forward, her notes became more specific, extensive, and erudite.

14. See, for instance, Pascoe, "Female Botanists," 199–202. Smith quotes the *Botanic Garden* on numerous occasions, and in a letter to her publishers on August 27, 1799, she praises it as "one of my favorite books" (*Collected Letters,* 332).

15. Mahood, *Poet as Botanist,* 20.

16. Labbe, "Transplanted," 72–73. George, *Botany, Sexuality,* 124.

17. Although Smith's letters show that she requests Aikin's text from her publishers in 1796, it is likely that she had read it earlier.

18. John Aikin, *Essay on the Application,* 1.

19. Ibid., 10, 1–2.

20. Ibid., 10.

21. Ibid., v.

22. Smith, *Poems,* 239; Smith, *Collected Letters,* 218.

23. Smith, *Poems,* 242, emphasis mine.

24. Smith, *Natural History of Birds,* 1:4.

25. Ibid., 2:91–93.

26. Smith, *Rural Walks,* 2:81.

27. Ibid., 2:81.

28. John Aikin, *Essay on the Application,* 132.

29. Smith, *Poems,* 200.

30. Ibid., 236.

31. Ibid., 239. Smith's further dispelling of general misconceptions can be seen, for instance, when she assures readers that the newt, though often "supposed to be venomous," is "perfectly harmless" (ibid., 302).

32. Mazzeo, *Plagiarism and Literary Property,* 185. Backscheider notices Smith's obsessive acknowledgments of borrowings: "Accused in her own time of plagiarism, she [Smith] responded by footnoting even the most familiar quotations and echoes in her verse" (*Eighteenth-Century Women Poets,* 335).

33. Smith, *Poems,* 66.

34. Smith, *Conversations,* 1:4.

35. Ibid., 1:46.

36. Ibid., 1:149, 179.

37. Porter, "From Nosegay to Specimen Cabinet," 29–44.

38. Smith, *Conversations,* 1:52, 1:56, 2:60.

39. Ibid., 2:ii. The text also includes a number of poems by Smith's "near relation," her sister, Catherine Dorset, with whom in 1797 Smith considered collaborating to write a botanical textbook; see Dolan, *Seeing Suffering,* 101.

40. Smith, *Conversations,* 1:96.

41. Smith, *Poems,* 107; Backscheider, *Eighteenth-Century Women Poets,* 336.

42. Mazzeo, *Plagiarism and Literary Property,* 2–5.

43. Smith, *Conversations,* 1:12.

44. Ibid., 1:179.

45. Ibid., 1:51.

46. Ibid., 2:65.

47. In regard to the collection of biological specimens, only collections of plants escape Smith's censure. For more on natural history collecting in the Romantic era, see, for example, Pascoe, *Hummingbird Cabinet;* and Tobin, *Duchess's Shells.*

48. Smith, *Conversations,* 2:64, 65.

49. Smith, *Poems,* 41.

50. Smith, *Conversations,* 1:51.

51. Smith, *Natural History of Birds,* 1:79.

52. For more on eighteenth-century conceptions of hybridity, see chapter 2.

53. Withering, "Introduction to the Study of Botany," 1:7. For more on botanists' views of hybrids and varieties, see George, *Botany, Sexuality,* 153–65.

54. Goldsmith, *History of the Earth,* 5:293.

55. Walker, *Elements of Geography,* 125.

56. Smith, *Natural History of Birds,* 2:29.

57. Smith, *Poems,* 260.

58. Smith, *Poems,* 303. Smith's distrust of such transformations accords with her adoption of Linnaean taxonomies and Linnaeus's early disregard for species alterations through hybridity; he wrote that "the number of species is constant" and "we count so many species as there were in the beginning" (Koerner, *Linnaeus,* 44). Conceiving of changes in the earth's distribution of populations, rather than within species themselves, Smith references extinctions of the walrus and a species of bird in Scotland, of wolves in Britain, and describes "elephant" remains found in England as well as in North America, along with those of the "rhinoceros and hippopotamus," and even predicts "the extirpation of the whole people [of Polynesia, particularly Tahiti]" (Smith, *Poems,* 294, 234, 245; *Natural History of Birds,* 2:21; *Rural Walks,* 147). However, these shifts in population remain consistent with species fixity as propounded, not only by Linnaeus, but also by the leading proponent of theories of extinction, Georges Cuvier. For Smith, although fanciers may produce varieties of species, each species fills its own connective place as an indelible "link" in nature so that, for instance, the fern owl "is a link between the Swallow and those birds that prey indiscriminately on smaller birds, insects, and reptiles," thus supporting the notion of an orderly, systematic configuration of species (Smith, *Natural History of Birds,* 2:118). Sounding like a modern environmentalist, and retaining this notion of "link[s]" between species, Smith additionally points out interdependencies that keep ecologies in a delicate balance and that humans can disrupt only with dire consequences, as when she relates that a species of bird was exterminated in a particular area, allowing a kind of caterpillar to reproduce unchecked and thus wreak destruction on local crops (ibid., 1:68–69).

59. Edward Young, *Conjectures,* 68.

60. Labbe, "Hybrid Poems," 221.

61. Fletcher, *Charlotte Smith,* 261.

62. Smith, *Conversations,* 2:v.

63. Ibid., 2:16.

64. For more on this poem, see Mellor, "Baffling Swallow," 299–309. Gilbert White also was very interested in debates about swallow migration or hibernation. For more on White, see Bellanca, *Daybooks of Discovery.*

65. Adam Smith, *Essays on Philosophical Subjects,* 13.

66. Smith, *Natural History of Birds,* 1:140.

67. Ibid., 2:60–61.

68. Pennant, *Literary Life,* 2, 5.

69. Koerner, *Linnaeus,* 113.

70. Mackay, "Agents of Empire," 39–40.

71. Smith, *Natural History of Birds,* 1:iii.

72. See Labbe, "'The absurdity of animals," 157–62.
73. Smith, *Natural History of Birds,* 1:60.
74. Ibid., 2:27.
75. Ibid., 2:64.
76. Darwin, *Zoonomia,* 154–55.
77. Seward, *Memoirs,* 217. Smith, *Letters,* 618.
78. Quoted in Mazzeo, *Plagiarism and Literary Property,* 44.
79. Smith, *Natural History of Birds,* 2:72–73.
80. Smith quotes from Darwin's chapter "On Instinct" in her section on the cuckoo in *Natural History of Birds* (1:73).
81. Darwin, *Zoonomia,* 155.
82. John Aikin, *Essay on the Application,* 136.
83. Ibid., 137–39.
84. Mazzeo, *Plagiarism and Literary Property,* 3.
85. Seward, *Letters,* 2:224. Seward mistakenly writes that the quotations and notes were added to the second edition of Smith's *Elegiac Sonnets* rather than the third.
86. "The Dictatorial Owl" also was omitted from *Beachy Head.*
87. Smith, *Poems,* 251.
88. John Aikin, *Essay on the Application,* 98, 100. Aikin also objects to fable because it "considers every animal as a human creature; and therefore has to do only with such of its qualities as bear a resemblance to the affections and manners of mankind, not with such as peculiarly constitute its natural history" (99).
89. In this statement, Smith may refer to Joseph Warton's essay that argues that "the works of those who profess an art whose essence is imitation, must needs be stamped with a close resemblance to each other; since the objects material or animate, extraneous or internal, which they all imitate, lie equally open to the observation of all, and are perfectly similar. Descriptions, therefore, that are faithful and just, must be uniform and alike: the first copier may be, perhaps, entitled to the praise of priority; but a succeeding one ought not certainly to be condemned for plagiarism" (*The Adventurer,* no. 63 [Tuesday, June 12, 1753]: 374). Another possible source may be William Derham's *Physico-Theology,* where he defends himself against similarities between his work and those of both John Ray and Robert Hooke, writing, "it is scarce possible, when Men write on the same, or a Subject near a-kin, and the observations are obvious, but that they must often hit upon the same thing: and frequently it happens from Persons making Observations about one and the same thing, without knowing what each other have done" (9).
90. John Aikin, *Essay on the Application,* 87.
91. Smith, *Natural History of Birds,* 2:45.
92. Wordsworth, "Essay, Supplementary to the Preface," in *William Wordsworth,* 650.
93. See Mazzeo, *Plagiarism and Literary Property,* 187. Also Macfarlane, *Original Copy*; and Stillinger, *Multiple Authorship.*
94. Peacock, *Peacock's Four Ages of Poetry,* 15–16. See Porter, "Formal Relocations," 672. Wordsworth and Coleridge, for example, both include notes to

some poems, referencing naturalists. Of particular interest to the present chapter, Coleridge displays what could be interpreted as authorial anxiety resembling that of Smith, denying a potential plagiarism, when, in his note to "This Lime Tree Bower My Prison," he explains, "Some months *after* I had written this line [about a rook], it gave me pleasure to find that [the naturalist, William] Bartram had observed the same circumstance of the Savanna Crane" (emphasis mine).

5. Translating Cosmopolitanism

1. Drouin and Bensaude-Vincent, "Nature for the People," 408

2. Burwick, "Romantic Theories of Translation," 69.

3. Ibid., 69.

4. Williams, introduction to *Tour in Switzerland,* ed. Vincent and Widmer-Schnyder, 14.

5. For a recent study of the complex dynamics of cosmopolitanism and globalism in this era, see Gottlieb, *Romantic Globalism.*

6. Fulford, Lee, and Kitson, *Literature, Science, and Exploration,* 4.

7. Ibid., 35, 9.

8. Ibid., 36.

9. Ibid., 10.

10. Glyn Williams, *Death of Captain Cook,* 40. This author explains that such portrayals of Cook contributed to the lionization of Cook's memory, but other contemporary accounts suggested Cook may have fallen victim to his "excessive anger" rather than excessive humanity.

11. Kippis, *Life of Captain James Cook;* my quotations are from a more recent reprint, *Captain Cook's Voyages,* 376.

12. Kippis, *Life of Captain James Cook,* 376.

13. "Edwin and Eltrada" (1782), "Ode on the Peace" (1783), and *Peru* (1784).

14. Gravil, "Helen Maria Williams," 57. Wordsworth would not actually meet Williams until 1820, but Gravil posits that "she was, more than Charlotte Smith or Felicia Hemans, his Sappho—his 'tenth muse.'"

15. Seward, *Letters,* 2:178.

16. Ibid., 2:220.

17. Seward's *Elegy on Captain Cook* is in *Poetical Works* 2: 33–46; the poem is cited by line number in this chapter's text.

18. Seward, *Poetical Works,* 2:40, 42.

19. Ibid., 2:41.

20. Ibid., 2:44–45.

21. Guest, *Small Change,* 253.

22. Williams's "The Morai" is in Kippis's *Life of Captain James Cook,* and is cited by line number in the text of this chapter.

23. Pratt, *Imperial Eyes,* 7.

24. Williams's mentor, Kippis, participated in the Revolution Society and viewed the natural sciences as a space in which "nations of the earth [can] subsist together in mutual agreement," for, "while the political world is always likely to be more or less the scene of altercation," the world of science "might be ex-

pected to continue in a state of tranquility and harmony" (Kennedy, *Helen Maria Williams*, 53; Kippis, *Considerations*, 3; Kippis, *Observations*, 2).

25. Kippis, *Life of Captain James Cook*, 351.

26. See Williams's poem "An Address to Poetry," ll. 89–96.

27. Williams, *Letters from France*, 1:1:14; henceforward, the volumes and page number of this work will be cited in the text of this chapter.

28. Bohls cites this quote to exemplify Williams's "affective connections to particular individuals rather than abstract or generalized reasoning" (*Women Travel Writers*, 123).

29. Kennedy, *Helen Maria Williams*, 111.

30. Williams, *Poems on Various Subjects*, 211; Williams, preface to Saint-Pierre, *Paul and Virginia*, vi.

31. Scientific advancements in the study of nature inspired Parisians to extirpate elaborate artificialities of, for instance, manner and dress associated with the *ancien régime*, and nature became representative of the revolution itself, as Williams witnessed in the planting of "the tree of liberty" near her lodgings, and in the republic's new calendar that renames months of the year to signify seasonal images in nature (*Letters from France*, 2:1:202–4, 1:2:195).

32. Williams, preface to Saint-Pierre, *Paul and Virginia*, iii.

33. Ibid., vii–viii.

34. Ibid., viii, x.

35. See also Kirkley, "Translating Rousseauism," 106–7.

36. Williams, *Tour in Switzerland*, ed. Vincent and Widmer-Schnyder, 24. When Williams escaped to Switzerland, she did not travel alone. At least as scandalous to British critics as Williams's political involvement in France were rumors of her relationship with the British reformer and entrepreneur John Hurford Stone. Seized letters from Stone to his brother, William, in Britain raised suspicion of treason, eventually resulting in William Stone's trial and acquittal in 1796; letters from Williams and J. H. Stone purportedly intercepted en route to Joseph Priestley in America added fuel to British anti-Jacobin propaganda by predicting a French invasion of England and cheerfully informing Priestley "that our OLD COUNTRY is now the only one left to struggle against the French Republic, and left under every disadvantage that every friend to her real welfare would wish" (Stone and Williams, *Copies of Original Letters*, 12–13). In addition to her patriotism, Williams's relationship with Stone called into question her virtue. Although Stone's wife divorced him in Paris earlier in 1794, rumors spread of an adulterous affair, exemplified in a gossiping letter from Hester Thrale Piozzi, once a friend of Williams, to Penelope Pennington on February 17, 1795: "The Rival Wits say that Helen Williams is turn'd *to Stone*, and tho' she was once Second to nobody, she is now Second to his Wife, who it seems was not guillotined as once was reported; but remains a living spectatress of the Political and *Im*politic Revolutions" (Piozzi-Pennington Letters at Princeton University; see Damián, "Helen Maria Williams's Personal Narrative," paragraph 5). Clearly sardonic in its intention, the pun on "Stone" figures Williams both as an adulteress and as exhibiting, as James Boswell accused, an unfeminine hardness of heart in supporting a revolution of such gruesome atrocities, thus denigrating her reputation as a poet of sensibility (Kennedy, *Helen Maria Williams*, 99–100).

37. Williams, *Tour in Switzerland* (1798); henceforward abbreviated *T* and cited by volume and page number in the text of this chapter.

38. Oldroyd, *Thinking about the Earth,* 134. However, as Heringman notes, there were certainly women interested in the earth sciences, and "Arguably the most fashionable woman in England, Georgiana, the Duchess of Devonshire, became a life-long devotee of mineralogy while on a continental tour in 1793," *Romantic Rocks,* 16.

39. Ramond's "Observations" supplemented his French translation of William Coxe's *Travels in Switzerland* (1779; trans. 1787): Coxe, *Lettres de M. William Coxe,* 2:100–148.

40. Bailey, *James Hutton.*

41. Darwin, *Botanic Garden* (1791), 43. Heringman discusses "the politicized Neptunist-Plutonist debate" and Darwin's and Hutton's theories of the earth, based on internal heat, in *Romantic Rocks,* 104–5.

42. Earlier in *Economy of Vegetation,* Darwin provides additional causes for the cooler climate in mountains, but these would not alter Williams's critique regarding the earth's ability to melt the glaciers.

43. Leask suggests that Williams's poem and translation of Ramond may have helped inspire Percy Shelley's poem *Mont Blanc* (1816); see "Mont Blanc's Mysterious Voice", 197, 201–2.

44. Darwin, *Botanic Garden* (1791), 54.

45. *Monthly Review* 27 (1798): 139.

46. Ibid., 140.

47. Ibid.

48. Ibid., 140, 144.

49. *European Magazine* 33 (1798) 391.

50. Buckland, in *Novel Science,* argues that opposition between these two schools of thought, catastrophism and uniformitarianism, was not so pronounced as many narratives of the history of geology would suggest.

51. Oldroyd, *Thinking about the Earth,* 135.

52. In the eighteenth century the term "revolution" could be problematic since, especially before the French Revolution, it could refer to either a cyclical or noncyclical notion of change, or even an ambiguous combination of the two. See Cohen, *Revolution in Science,* 208–9.

53. Quoted in ibid., 206.

54. Ramond makes this claim in his translation of Coxe; see Bewell, *Wordsworth and the Enlightenment,* 246.

55. Quoted in Cohen, *Revolution in Science,* 206.

56. For Ramond, the Alps are significant largely because he views them as exempt from change subsequent to their initial formation. In this he references "Saussure, [who,] in his excellent work on the Alps . . . has discovered, or thinks he has discovered, regular strata, in the primitive mountains; which consequently cannot be the work of a revolution" (Williams, *T,* fn. 2:287–88).

57. Ramond acknowledges that Pierre-Michel Hennin measured the advancement of "the Glaciers of Faucigny" as fourteen feet a year, and asserts that this "enormous" rate is far from universal, enumerating the various factors that may modify a glacier's progress; see Williams, *T,* 2:312. For more on the era's

interest in "primitive" and "secondary rocks," see Heringman, *Romantic Rocks*, 109–12.

58. See, for instance, Bewell, "Jefferson's Thermometer," 111–38.

59. For insight into the significance of volcano imagery during the French Revolution, see Miller, "Mountain, Become a Volcano," 555–585.

60. For a more thorough treatment of Buffon's theory, see Oldroyd, *Thinking about the Earth*, 91.

61. Percy Shelley, *Complete Works*, 6:139.

62. Shelley may differ from Williams in that he seems to reject Saussure's assertion (largely adopted by Ramond) that glaciers "have their periods of increase and decay," insisting instead on glaciers' "slow but irresistible progress" (ibid., 6:139).

63. Montesquieu, *Spirit of the Laws*.

64. Within *Essay on the Geography of Plants*, Humboldt calls for a physical tableau of European vegetation in the order of that he has created for the equatorial regions and declares Ramond the ideal naturalist to carry out such a task. As Williams had done in her work, Humboldt alludes to his private correspondence with Ramond, whom he finds "equally knowledgeable in geology and botany, possess[ing] both the art of observing well and the talent of speaking to the imagination" (Humboldt and Bonpland, *Essay on the Geography of Plants*, 94). In his botanical geography of the Americas, Humboldt, like Ramond, creates overt social analogies, used by the younger naturalist to comment on both governmental and scientific institutions.

65. Humboldt and Bonpland, *Essay on the Geography of Plants*, 4.

66. Humboldt, *Personal Narrative* (1818, 1966), vi; all quotes from this text are from volume 1. The botanist Aimé Bonpland shared Humboldt's exploration, but Humboldt received the lion's share of the glory.

67. Dettelbach, "Humboldtian Science," 289. Cook's voyages shaped Humboldt's scientific drive on a personal level, for in 1790 Humboldt traveled from Germany to England and revolutionary Paris with Georg Forster, who had been a naturalist on Cook's second voyage (1772–75), along with his father, Johann Reinhold Forster. The elder Forster promoted vegetation as the environmental aspect most directly affecting humanity, as the mediating component between physical phenomena (such as climate) and humans, a concept that influenced Humboldt's thinking. Additionally, studying under a leading proponent of Neptunism, Abraham Werner, at the Freiberg School of Mines in 1791, Humboldt absorbed Werner's critiques of contemporary geology as tending to be too speculative. Werner coined the term "geognosy" to distinguish his new scientific program that was to be firmly grounded in facts. Although Humboldt would later convert to Vulcanism, geognosy, as a type of new historical geology, influenced his program for plant geography, which required botany to give insight into the history of the earth. See Dettelbach, "Global Physics and Aesthetic Empire," 267.

68. Nicolson, introduction, xviii.

69. Dettelbach, "Humboldtian Science," 299.

70. Humboldt, *Personal Narrative* (1818, 1966), vi–vii.

71. Qtd in Leask, *Curiosity and Aesthetics,* 290.

72. Jason Wilson, translator's preface to Humboldt, *Personal Narrative* (1995), lix.

73. Ibid.

74. Humboldt, *Personal Narrative* (1818, 1966), xi.

75. Qtd. in Wilson, translator's preface to Humboldt, *Personal Narrative* (1995), lx.

76. For Humboldt, imagination and feeling "could, if suitably trained and applied, transcend the limitation of reason, penetrate beyond surface phenomena and, sensuously and intuitively, grasp the underlying unities of nature" (Nicolson, introduction, xx).

77. Humboldt and Bonpland, *Essay on the Geography of Plants,* 31.

78. Ibid., 61.

79. Ibid., 84.

80. Ibid., 65.

81. Ibid., 133.

82. Pérez-Mejía, *Geography of Hard Times,* 71. Identifying Humboldtian science as a revolution, Louis Agassiz enthused that Humboldt's American travels "completely changed the basis of physical sciences as the revolution which took place in France about the same time has changed the social condition of that land" (qtd. in Stephen T. Jackson, introduction to Humboldt and Bonpland, *Essay on the Geography of Plants*, 9). In 1821 Simón Bolívar, the anticolonialist leader and president of the newly declared country of Gran Colombia, called Humboldt "the new discoverer of America" and stated that "he has done more for America than all the conquistadors together" (qtd. in Pérez-Mejía *Geography of Hard Times*, 40). Pérez-Mejía, however, qualifies that "what Bolívar meant, and what is still echoed today, is that he led us out of 'barbarity' and toward the light of Europe" (58). She argues that this is problematic because Humboldt's maps and knowledge of nature largely derive from information gleaned from native people of the lower classes who served as the traveler's guides and are themselves made "analyzable and classifiable" within his work. The convictions of the French Revolution and Humboldt's Enlightenment liberalism mask the occurrence of certain indigenous forms of oppression while altering relations with Europe, for throughout the colonies, native people "were fighting for their lands against the creoles, but in Humboldt's text their political agency is forgotten" (70).

83. Humboldt, *Personal Narrative* (1818, 1966), 1.

84. Ibid., 2.

85. Ibid.

86. Ibid.

87. Leask writes that "in the very act of translation, Williams appropriated Humboldt's text for non-specialist and women readers" ("Salons, Alps, and Cordilleras," 225).

88. Martin, "These Changes and Accessions of Knowledge," 46.

89. Humboldt, *Personal Narrative* (1818, 1966), vii–viii.

90. Comparisons between Humboldt and Napoleon flourished in the nineteenth century. Both men were born in 1769, and while Humboldt's quest for universal domination of nature generally appealed more to public sentiment

than Napoleon's quest for universal domination of nations, the strength of each man's personality and global ambition prompted correlation. The *Encyclopedia Britannica* claimed for many editions that Humboldt's fame in Europe was second only to that of Napoleon (Sachs, *Humboldt Current*, 380n17). If Napoleon could crown himself Emperor, Humboldt seemed, in Harriet Martineau's words, the self-made "Monarch of science" (qtd. in Wilson, translator's preface to Humboldt, *Personal Narrative* [1995], xxviii). Emerson called Humboldt "the Napoleon of travelers" who "marches like an army, gathering all things as he goes," and another American poet, Henry T. Tuckerman, dubbed Humboldt "the Napoleon of science" (see Walls, *Passage to Cosmos*, 14, 254–55). Oliver Wendell Holmes compares the two men in his poetic tribute for the centennial of Humboldt's birth, "Humboldt's Birthday: Bonaparte, Aug. 15th, 1769—Humboldt, Sept. 14th, 1769," describing the naturalist as a "peaceful conqueror" achieving "bloodless triumphs" (see Sachs, *Humboldt Current*, 380n17; Walls, *Passage to Cosmos*, 310).

91. Pratt, *Imperial Eyes,* 124.

92. Williams, *Poems on Various Subjects,* xv.

93. Ibid.

94. Drouin and Bensaude-Vincent, "Nature for the People," 411.

95. Pratt, *Imperial Eyes,* 117; Dettelbach, "Global Physics," 283.

96. Napoleon stated, "So, monsieur, you collect plants? So does my wife." See Sachs, *Humboldt Current,* 38.

97. See, for instance, Marina Benjamin's classic studies, *Science and Sensibility,* and Benjamin, *Question of Identity.*

98. Williams, *Poems on Various Subjects,* 155.

99. See Mellor, "Possessing Nature," 220–32; and Mellor, "Frankenstein: A Feminist Critique of Science," 287–312. For further discussion about how ideas of hybridity and reproducibility relate to female fertility, obstetrics, and authorship, see, for example, Gilbert and Gubar, *Madwoman in the Attic;* Todd, *Imagining Monsters;* Cody, *Birthing the Nation;* and Mellor, *Mothers of the Nation.*

6. Reconstructing Origins

1. Shelley, *The Last Man,* 340. Hereafter abbreviated *L* and cited parenthetically by page number in the text of this chapter.

2. See, for instance, Sophie Thomas, "Ends of the Fragment," 22.

3. See Stafford, *Last of the Race.*

4. For thought-provoking discussions of the forms of science operating in *Frankenstein,* see especially, Vasbinder, *Scientific Attitudes;* Butler, "Frankenstein and Radical Science," 302–13; Mellor, "Feminist Critique of Science," 62–87. For essays addressing issues of empire, political revolution, gender, and/or disease in *The Last Man,* see, for example, Mellor, *Mary Shelley;* Cantor, "Apocalypse of Empire," 193–211; Lew, "Plague of Imperial Desire," 261–78; Bewell, *Romanticism and Colonial Disease;* Lokke, "*The Last Man,*" 116–34.

5. Shelley, *Journals,* 476–77.

6. Shelley, *Letters,* 1:566.

7. See, for instance, Lokke, "*The Last Man,*" 128.

8. Percy Shelley, *Letters of Percy Bysshe Shelley,* 2:458–59.

9. Rudwick, *Georges Cuvier,* 183. For an excellent study of the relationship between natural history and antiquarianism in this era, see Heringman, *Sciences of Antiquity.*

10. Cuvier, "Memoir on the Species of Elephants," was read at the public session of the National Institute on April 4, 1796, and later published in the institute's collection. See Rudwick, *Georges Cuvier,* 18–19.

11. Rudwick, *Georges Cuvier,* 24.

12. Bowler, *Evolution,* 111.

13. Bennett made a similar comparison when she wrote, "In *Frankenstein,* Mary Shelley constructed a being from parts to startle the world from its complacency; in *The Last Man,* she again constructs from parts, this time, the written word" ("Radical Imaginings," 152).

14. Report of an 1834 lecture, qtd. in O'Connor, "Byron's Afterlife," 154.

15. *Monthly Review,* n.s. 1 (March 1826): 335, emphasis mine.

16. Byron, *Lord Byron,* 882. *Cain* is hereafter abbreviated *C* and cited parenthetically by act, scene, and line number. In Britain, Cuvier's "Preliminary Discourse" was quickly translated into English in 1813 by Robert Jameson, who added a preface and additional notes, as well as a new title, to make Cuvier's work align more closely with biblical scripture. In particular, Jameson's highly successful editions of Cuvier's text associated the earth's most recent revolution with Noah's Flood, a suggestion eagerly accepted by many British geologists, though not Cuvier's intention. Much more than on the continent, British geologists felt pressure to interpret their studies of the earth in ways that would uphold scriptural authority. See Karkoulis, "They Pluck'd the Tree of Science," 273–81; and O'Connor, "Mammoths and Maggots," 26–42.

17. Shelley, *Letters,* 1:209; Styles, *Lord Byron's Works,* 3, 15. The *OED* defines the phrase "deep time" as "Time in the far distant past or future; time viewed on a geological or cosmological scale rather than the historical scale" and attributes the earliest usage to Thomas Carlyle in 1832, though John McPhee is often credited with bringing the phrase into popular usage with his book *Basin and Range.*

18. For the full text of this poem, see O'Connor, "Kirkdale Cave," 39–41.

19. Buckland, "The Professor's Descent," ll. 35–40.

20. See Jeffrey, "Cuvierian Catastrophism," 150. Heringman suggests that in *Prometheus Unbound,* "Shelley's account of animal remains . . . is the most visibly indebted to Parkinson and geologists such as Cuvier, Smith, and Buckland, who revived catastrophism and generated the fashion (roughly 1815–1830) for Deluges and dinosaurs" ("Rock Record," 72).

21. Buckland, "The Professor's Descent," ll. 47–51.

22. Rupke, *Great Chain of History,* 7.

23. Ibid., 64.

24. Although this contradicted Cuvier's idea of continents and oceans exchanging places, the French naturalist sanctioned Buckland's hyena-den theory, perhaps largely because Buckland's cave work did corroborate Cuvier's notion of successive, cataclysmic revolutions marking divisions within the earth's history.

25. Conybeare's poem, ll. 17, 44. Rudwick, *Scenes,* 39; Sommer, *Bones and Ochre,* 48; O'Connor, *Earth on Show,* 95.

26. Rudwick, *Scenes,* 39.

27. Daubeny, *Fugitive Poems,* 92–94; the poem is also quoted in Rudwick, *Scenes,* 40–43.

28. Sommer, *Bones and Ochre,* 49.

29. Shelley, *Letters,* 1:212.

30. Rudwick, *Georges Cuvier,* 182.

31. In employing this goat skeleton as a symbol of catastrophe, Shelley could not have known of a comical incident in 1826 in which Buckland informed the population of Palermo that bones found in a cave and believed to be those of Rosalia, the city's patron saint, in fact belonged to a goat, Rupke, *Great Chain of History,* 68.

32. Shelley, *Journals,* 423.

33. Byron, *Don Juan,* 9.291, in *Lord Byron,* 305–7, 320. See also Jones, "When this World Shall Be Former."

34. Rudwick, *Georges Cuvier,* 18, 257.

35. Williams, *Tour in Switzerland,* 2:347.

36. In detailing this astronomical anomaly, Shelley appeals to biblical literalism even as she makes no direct mention of supernatural forces. The three meteors arguably contain the scriptural significance of the trinity, especially as "united" in a single entity, and the resulting tidal wave threatens forfeiture of God's biblical promise never again to destroy the world with flood. While the depiction of cattle plunging over the cliff's edge to their deaths in the ocean seems to reference the swine herd that runs off a precipice and falls into the sea in the gospels at the instigation of demons, Shelley's cattle race to their death in terrified reaction to the inexplicable, but not obviously supernatural, three whirling "suns" (Genesis 9:8–11; Matthew 8:28–34; Mark 5:1–20; Luke 8:26–39).

37. See Rupke, *Great Chain of History,* 41, 76–77.

38. Shelley, *Prometheus Unbound,* act 4, ll. 309, 310, 316–18.

39. Cameron, "Mary Shelley's Malthusian Objections," 177–203. For texts referencing the hypothesis that dinosaurs became extinct due to plague, see, for instance, Gould, *Hen's Teeth,* 320, and Gallagher, *When Dinosaurs Roamed,* 114. Of course the term "dinosaur" was not coined until 1842, by Richard Owen.

40. As Webb puts it, "The collapse of the domestic world prefigures the collapse of the entire human world" ("Reading the End of the World," 121).

41. Bewell, *Romanticism and Colonial Disease,* 306; Cantor, "Apocalypse of Empire," 196.

42. Stafford, *Last of the Race,* 216.

43. Near the beginning of the eighteenth century, Sir Isaac Newton, for instance, estimated the creation of the earth to have occurred in 4000 BC; a century later, Cuvier estimated the earth to be millions of years old, but other catastrophists maintained estimates closer to that of Newton, taking a literal interpretation of the "days" of creation in Genesis. James Hutton, on the other hand, claimed that the earth's history stretched indefinitely into the distant past, helping to form the ideas of uniformitarianism and of deep time.

44. Abrams, *Natural Supernaturalism,* 12. Mellor also notices that Shelley departs from the Wordsworthian "paradigm of natural supernaturalism" (*Mary Shelley*, 165).

45. Abrams, *Natural Supernaturalism,* 47, 334.

46. Ibid., 338, 96.

47. It seems no coincidence that Shelley references Haydn's song "New-Created World" (446).

48. Abrams, *Natural Supernaturalism,* 12, 48.

49. Quoted by Rudwick, *Scenes,* 48.

50. See Kunzle, "World Upside Down," 39–94.

51. Bewell suggests the possibility of inoculation in *Romanticism and Colonial Disease,* 313.

52. In this vein, Sussman writes, "The creature in *Frankenstein* is monstrous because society deems him so; Verney is monstrous because society has disappeared around him" ("Islanded in the World," 298).

53. Duncan, "The Last Hyæna," ll. 11–16, in Daubeny, *Fugitive Poems,* 119–20.

54. Ibid, ll. 27–30.

55. O'Connor, *Earth on Show,* 43–44.

56. Spenser, *Ruins of Rome* (1591), in *Complete Works,* l. 406.

57. See Mellor, *Mary Shelley,* 169. My reading, emphasizing both the mind and this search for meaning, in this sense, perhaps falls somewhere between the interpretations of Mellor and Fiona Stafford. The latter writes that Mellor's "nihilistic reading is contradicted by the text itself, which repeatedly emphasizes not the meaningless[ness] of the human mind, but its supreme value" (*Last of the Race,* 226).

Conclusion

1. As Scheinberg argues, Hemans and other Victorian women writers rejected the model of "poet as prophet" so crucial to the male tradition of Romanticism, and adopted instead a poetic identity rooted in an understanding of "poetry as theology" in which women could claim "Christian poetic authority through the discourse of the heart" (*Women's Poetry,* 51).

2. *British Critic,* n.s. 20 (July 1823): 50–61, qtd. in Wolfson, *Felicia Hemans,* 537.

3. Kelly, "Death and the Matron," 196.

4. Chorley, *Memorials,* 1:187–88.

5. Gordon, *Life and Correspondence,* 123. For more on the various institutions that developed during this time, see Klancher, *Transfiguring.*

6. Chorley, *Memorials,* 1:47. Kelly also discusses Hemans's decision not to publish "overtly satirico-political verse," such as "The Army" and "Reform," because they "would have been considered by many to be unfeminine" (*Felicia Hemans,* 21). Three years later after Chorley's publication, Hemans's sister, Harriet Browne Owen, under the pseudonym of Mrs. Hughes, wrote a second biography of Hemans's life, including this epitaph and a related, previously unknown poem, Hemans's "Epitaph on the Hammer of the Aforesaid Mineralogist," assuring that, "as may easily be supposed, [these poems] were never intended for publication, but were merely a *jeu d'esprit* of the moment" (Owen, *Memoir,* 46).

7. Harriet Browne Owen, letter, National Library of Scotland MS 4090, folio 193–94.

8. This poem can be found in Wolfson, *Felicia Hemans.*

9. This describes an area in Wales.

10. Many contemporary geological sites referenced Lucifer, such as Derbyshire's Devil's Peak (also known as Devil's Cave, Devil's Bottom, and Devil's Arse). For further discussion of Devil's Cave, see Heringman, *Romantic Rocks,* 245–51.

11. Wilton, *Geology and Other Poems.*

12. Smith, *Poems,* 280.

13. Wolfson, *Felicia Hemans,* 16. For example, in Charlotte Smith's "Flora," cited above, she draws on mock-heroic conventions found in Alexander Pope's *The Rape of the Lock.*

14. See O'Connor, *Earth on Show,* 81–85, for further discussion of this fascinating corpus of geological verse.

15. Whately, "Elegy Intended for Professor Buckland," ll. 21–26, 33–38, in Daubeny, *Fugitive Poems.*

16. See De Carolis and Patricelli, *Vesuvius,* 77–82.

17. Wolfson, *Felicia Hemans,* 424.

18. De Carolis and Patricelli, *Vesuvius,* 102.

19. Ibid., 118.

20. Armstrong, "Natural and National Monuments," 218.

21. Dwyer, *Pompeii's Living Statues,* 1.

22. "Image in Lava" can be found in Wolfson, *Felicia Hemans.*

23. Melnyk, "William Wordsworth," 143–44.

24. Leighton, *Victorian Women Poets,* 21.

25. Lundeen, for example, explores similarities between these two poems in her article "When Life Becomes Art."

26. Armstrong reads this line similarly in "Natural and National Monuments," 228.

27. Leighton, *Victorian Women Poets,* 2, 3.

28. Wyatt, *Wordsworth and the Geologists,* 19.

29. "Resolution and Independence" can be found in Wordsworth, *William Wordsworth;* lines from this poem are cited in the text of this chapter.

30. Bewell, *Wordsworth and the Enlightenment,* 265; Heringman, *Romantic Rocks,* 40.

31. Bewell, *Wordsworth and the Enlightenment,* 266.

32. Abrams, "Structure and Style," 545, quoting from *The Ruined Cottage* (1797–98).

33. Thomas and Ober, *Mind For Ever Voyaging,* 4, 5.

34. Hemans published her poem "To Wordsworth" in the *Literary Magnet* in 1826, visited the older poet in the summer of 1830, and dedicated to him her verse volume *Scenes and Hymns of Life; with other Religious Poems* (1834).

35. Melnyk, "William Wordsworth," 140.

36. Ibid., 145.

37. "Wood Walk and Hymn" is from Hemans, *Scenes and Hymns of Life;* lines from this poem are cited in the text of this chapter.

38. Murphy, *In Science's Shadow*, 1.
39. See chapter 1.
40. Brooks, "Condition of Women," 154–55.
41. Murphy, *In Science's Shadow*, 23.
42. Huxley, *Science and Education*, 71; Pike, "Woman and Political Power," 86.
43. Battersby, *Gender and Genius*, qtd in Peterson, *Becoming a Woman of Letters*, 44.
44. Heráud, "On Poetical Genius," 59, 63.
45. Hazlitt, *Lectures on the English Poets*, 318, 319–20.
46. Coleridge, *Biographia Literaria*, chap. 22.
47. Peterson, *Becoming a Woman of Letters*, 14.
48. Martineau, *Autobiography*, 126. Indeed, as Looser points out regarding eighteenth-century women writers, "A generation or two later, these once-dominant writers had become difficult to trace. It seems mindboggling that by the mid-nineteenth century, Elizabeth Barrett Browning could not locate any literary 'grandmothers' among British poets" ("Why I'm Still Writing Women's Literary History," 224).
49. Pettitt, *Patent Inventions*, 36–83.
50. Frank Turner, *Contesting Cultural Authority*, 171–200.
51. Peterson, *Becoming a Woman of Letters*, 35.
52. Macfarlane, *Original Copy*, 39.
53. Abrams, *Mirror and the Lamp*, 24.
54. Carlyle, *Works of Thomas Carlyle*, 5:45.
55. For an excellent essay examining changes in the relationship of science and religion in the western world between 1750 and 1870, see Frank Turner, "Late Victorian Conflict," 87–110.
56. See also Menke, "Political Economy of Fruit," 105–36.
57. Eliot, *Impressions of Theophrastus Such*, 58.
58. Pearson, *Women's Reading in Britain*, 67; O'Connor, *Earth on Show*, 4.
59. Kucich, "Scientific Ascendancy," 121.
60. Ibid., 123. See also Shuttleworth, *George Eliot*.
61. Kucich, "Scientific Ascendancy," 134.
62. See Gates, *Kindred Nature*; and Sheffield, *Revealing New Worlds*.
63. Barbauld, *Pleasures of Imagination*, 2.
64. Ibid., 4.
65. Ibid.
66. Ibid., 5.
67. Perhaps the most famous example of this public satire is George Canning's scathing parody of Erasmus Darwin's *The Loves of the Plants*, titled, *The Loves of the Triangles* (1798), published in *The Anti-Jacobin Review*.
68. Southey, *Omniana*, 1:101. My thanks to Allison Cooper Davis for bringing this quote to my attention.
69. See Shteir, *Cultivating Women*, 149–69.
70. Huxley, *Life and Letters*, 2:337. As scientific ideology exerted influence on literary practice, so did literature continue to shape science as well. Beer explains, "The organization of *The Origin of Species* seems to owe a good deal to . . . Charles Dickens, with its apparently unruly superfluity of material gradually

and retrospectively revealing itself as order" (*Darwin's Plots*, 6). In a letter to Hallam Tennyson, Tyndall also praised Alfred Tennyson's scientific studies: "In regard to metaphors drawn from science, your father, like Carlyle, made sure of their truth. To secure accuracy, he spared no pains" (Tennyson, *Alfred, Lord Tennyson*, 2:475). In the 1980s such quotes provided evidence for literary scholars such as Cosslett, who, while admitting that "Tennyson also had many reservations about the tendencies of contemporary science," sought to overturn the notion that nineteenth-century "science and literature (or art) are utterly independent, mutually antagonistic modes of thought, 'two cultures', as the popular phrase (of C. P. Snow) has it" (*The "Scientific Movement"*, 39; Dale, *In Pursuit of a Scientific Culture*, 8; see also Chapple, *Science and Literature*). In addition to Dickens and Tennyson, these scholars often highlight science in the literature of George Eliot, George Meredith, Gerard Manley Hopkins, and Thomas Hardy. More recently, Dawson cautions that the now prevalent "One Culture" model of literature and science scholarship, popularized by critics such as Beer, Cosslett, and Levine, "has been much too sanguine in its approach to the interrelations of science and literature in the Victorian period" (*Darwin, Literature, and Victorian Respectability*, 7; Levine, *Darwin and the Novelists*). Revealing recurrent correlations between Darwin and sexual immorality in the era's print culture, Dawson argues that Huxley and Tyndall emphasized connections with reputable writers such as Tennyson and Carlyle, misinterpreting or reinterpreting these authors' works to bolster their own respectability and evolutionary science's cultural authority, Dawson, *Darwin, Literature, and Victorian Respectability*, 4, 21, 23.

BIBLIOGRAPHY

✝

Primary Sources

Addison, Joseph. *Spectator* 253 (Dec. 20, 1711).

Aikin, John. *An Essay on the Application of Natural History to Poetry.* Warrington: printed for J. Johnson, 1777.

Aikin, John, Sr. *An Essay on the Plan and Character of Thomson's Seasons.* London, 1788.

———. *The Calendar of Nature, or, Youth's Delightful Companion.* London, 1789.

Aikin, John, and Anna Barbauld. *Evenings at Home, or, The Juvenile Budget Opened.* 6 vols. London, 1792–96.

Aikin, Lucy, ed. *The Works of Anna Letitia Barbauld. With a Memoir.* 2 vols. London, 1825.

Barbauld, Anna. *The Pleasures of Imagination by Mark Akenside, M.D. to which is prefixed a Critical Essay on the Poem, by Mrs. Barbauld.* London, 1795.

———. "Prefatory Essay." *The Poetical Works of Mr. William Collins.* London, 1797. iii–xlix.

———. *The Poems of Anna Letitia Barbauld.* Ed. William McCarthy and Elizabeth Kraft. Athens: U of Georgia P, 1994.

———. *Anna Letitia Barbauld: Selected Poetry and Prose.* Ed. William McCarthy and Elizabeth Kraft. Toronto: Broadview P, 2002.

Born, Edler von Ignaz. *John Physiophilus's Specimen of the Natural History of the Various Orders of Monks, After the Manner of the Linnaean System, Translated from the Latin, Printed at Augsburgh.* London, 1783.

Brooks, W. K. "The Condition of Women from a Zoological Point of View I." *Popular Science Monthly* 15 (1879): 145–55.

Buffon, Comte de, Georges-Louis Leclerc. *Natural History, General and Particular.* 9 vols. Trans. William Smellie. Edinburgh: printed for William Creech, 1780–85.

Byron, George Gordon, Lord. *Lord Byron: The Major Works.* Ed. Jerome J. McGann. New York: Oxford UP, 1986.

Carlyle, Thomas. *The Works of Thomas Carlyle.* 30 vols. Ed. H. D. Traill. London: Chapman and Hall, 1896–99.

Chorley, Henry F. *Memorials of Mrs. Hemans with Illustrations of her Literary Character from her Private Correspondence.* 2 vols. London: Saunders and Otley, 1836.

Coleridge, Samuel Taylor. *Biographia Literaria.* 2 vols. London: Rest Fenner, 1817.

Coxe, William. *Lettres de M. William Coxe Am.W. Melmoth, sur L'Etat Politique, Civil et Naturel de la Suisse; Traduites de L'Anglois, et Augmentees des Observations Faites dans le meme pays, par le Traducteur.* Trans. Louis-François Ramond. 2 vols. Paris, 1787.

Crèvecoeur, J. Hector St. John de. *Letters from an American Farmer.* 1782. Ed. Albert E. Stone. New York: Penguin, 1986.

Cuvier, Georges. "Memoir on the Species of Elephants, Both Living and Fossil." *Magasin encyclopédique,* Apr. 4, 1796.

Darwin, Erasmus. *The Botanic Garden, Part II. Containing The Loves of the Plants, A Poem with Philosophical Notes.* Lichfield, 1789.

———. *The Botanic Garden; A Poem in Two Parts. Part I Containing the Economy of Vegetation. Part II. The Loves of the Plants. With Philosophical Notes.* London, 1791.

———. *Zoonomia; or the Laws of Organic Life.* Vol. 1. London: printed for J. Johnson, 1794–96.

Daubeny, Charles. *Fugitive Poems connected with Natural History and Physical Science.* Oxford: James Parker and Co., 1869.

Derham, William. *Physico-Theology.* 1713. 2nd ed. London: printed for W. Innys, 1714.

Duff, William. *An Essay on Original Genius; And its Various Modes of Exertion in Philosophy and the Fine Arts, Particularly in Poetry.* London: printed for Edward and Charles Dilly, 1767.

Edgeworth, Richard Lovell. *Memoirs of Richard Lovell Edgeworth, Begun by Himself and Concluded by His Daughter, Maria Edgeworth.* 2 vols. London: R. Hunter, 1820.

Edwards, Bryan. *The History, Civil and Commercial, of the British Colonies in the West Indies.* 2 vols. London, 1793.

Eliot, George. *Impressions of Theophrastus Such.* 1879. Ed. Nancy Henry. London: William Pickering, 1994.

Goldsmith, Oliver. *An Inquiry into the Present State of Polite Learning in Europe* London: printed for R. and J. Dodsley, 1759.

———. *An History of the Earth and Animated Nature.* 8 vols. London, 1779.

Grainger, James. *The Sugar-Cane: A Poem, in Four Books, with notes.* London: printed for R. and J. Dodsley, 1764.

Haywood, Eliza. *The Female Spectator.* 4 vols. London: printed and published by T. Gardner, 1745–46.

Hazlitt, William. *Lectures on the English Poets, Delivered at the Surrey Institution.* London: Taylor and Hessey, 1818.

Hemans, Felicia. *Scenes and Hymns of Life, with other Religious Poems.* Edinburgh: William Blackwood; London: T. Cadell, 1834.

Heráud, J. A. "On Poetical Genius Considered as a Creative Power." *Fraser's Magazine* 1 (1830): 56–63.

Heron, Robert (John Pinkerton). *Letters of Literature.* London: printed for G. G. J. and J. Robinson, 1785.

Humboldt, Alexander von. *Personal Narrative of Travels to the Equinoctial Regions of the New Continent, During the Years 1799-1804, by Alexander de Humboldt, and Aime Bonpland . . .* 1818. Trans. Helen Maria Williams. 7 vols. in 6. New York: AMS P, 1966.

———. *Personal Narrative.* 1814–29. Preface and trans. Jason Wilson. London: Penguin, 1995.

Humboldt, Alexander von, and Aimé Bonpland. *Essay on the Geography of Plants.* 1807. Ed. with an introduction by Stephen T. Jackson. Trans. Sylvie Romanowski. Chicago: U Chicago P, 2008.

Huxley, Thomas H. *Science and Education Essays*. New York: Greenwood, 1968.
———. *Life and Letters of Thomas Huxley*. 2 vols. Ed. Leonard Huxley Macmillan, 1990.
Johnson, Samuel. *Adventurer* 95, Oct. 2, 1753, 145.
———. *Idler,* no. 97, Saturday, Feb. 23, 1760.
Keats, John. *Letters of John Keats*. Ed. Robert Gittings. Oxford, New York: Oxford UP, 1970.
Kerr, Robert. *Memoirs of the Life, Writings, and Correspondence of William Smellie*. 2 vols. Edinburgh, 1811.
Kippis, Andrew. *Considerations on the Provisional Treaty with America, and the Preliminary Articles of Peace with France and Spain*. London, 1783.
———. *Observations on the Late Contests in the Royal Society*. London, 178-.
———. *Life of Captain James Cook*. London, 1788.
———. *Captain Cook's Voyages, with an Account of his Life*. New York: A. A. Knopf, 1925.
Lamb, Charles. *The Letters of Charles and Mary Anne Lamb*. 3 vols. Ed. Edwin W. Marrs Jr. Ithaca: Cornell UP, 1976.
Lennox, Charlotte. *Lady's Museum*. 2 vols. London: printed for J. Newbery, 1760–61.
Locke, John. *Two Treatises of Government*. 1690. Ed. Peter Laslett. Cambridge: Cambridge UP, 2002.
Long, Edward. *The History of Jamaica, or General Survey of the Antient and Modern State of that Island with Reflections on its Situation, Settlements, Inhabitants, Climate, Products, Commerce, Laws, and Government*. 3 vols. London: printed for T. Lowndes, 1774.
Martineau, Harriet. *Autobiography*. Ed. Linda H. Peterson. Petersborough: Broadview P, 2007.
Martyn, John. *The Georgicks of Virgil, with an English translation and notes by John Martyn, Professor of Botany at the University of Cambridge*. 3rd ed.. London, 1755.
Montesquieu, Baron de, Charles de Secondat. *The Spirit of the Laws*. 1748. Trans. Thomas Nugent. London: printed for J. Nourse and P. Vaillan, 1750.
More, Hannah. *Strictures on the Modern System of Female Education. With a View of the Principles and Conduct Prevalent among Women of Rank and Fortune*. 2 vols, 1799.
Owen, Harriet Browne. *Memoir of the Life and Writings of Mrs. Hemans*. Philadelphia: Lea and Blanchard, 1839.
Peacock, Thomas Love. *Peacock's Four Ages of Poetry, Shelley's Defence of Poetry, Browning's Essay on Shelley*. Ed. H. F. B. Brett-Smith. Boston, New York Houghton Mifflin, 1921.
Pennant, Thomas. *British Zoology*. 4 vols. Warrington, London: printed for Benjamin White, 1768–70.
———. *A Tour of Scotland, and Voyage to the Hebrides*. Chester: printed for John Monk, 1774.
———. *A Tour in Wales*. London: printed for Henry Hughes, 1778–83.
———. *History of Quadrupeds*. 2 vols. London: printed for Benjamin White, 1781.
———. *The Journey from Chester to London*. London: printed for Benjamin White, 1782.
———. *Of London*. London: printed for Robert Faulder, 1790.

———. *The Literary Life of the Late Thomas Pennant, Esq., By Himself.* London: printed for Benjamin White, 1793.

Percival, Thomas. *Moral and Literary Dissertations.* London: printed for J. Johnson, 1784.

Perry, James. *Mimosa: or the Sensitive Plant; A Poem dedicated to Mr. Banks.* London: printed for W. Sandwich, 1779.

Pike, Luke Owen. "Woman and Political Power." *Popular Science Magazine* 1 (1872): 82–94.

Ray, John. *Wisdom of God Manifested in the works of the Creation, in two parts.* London: printed for D. Williams, 1691, 1762.

Riddell, Maria. *Voyages to the Madeira, and Leeward Caribbean Isles: with Sketches of the Natural History of these Islands.* Edinburgh: printed for Peter Hill, 1792, 1802.

———, ed. *The Metrical Miscellany: Consisting Chiefly of Poems Hitherto Unpublished.* London: printed for Cadell and Davies, 1802.

Saint-Pierre, J. H. Bernardin. *Paul and Virginia.* Preface and trans. Helen Maria Williams. London: G. G. and J. Robinson, 1795.

Schaw, Janet. *Journal of a Lady of Quality; Being a Narrative of a Journey from Scotland to the West Indies, North Carolina, and Portugal, in the years 1774 to 1776.* Ed. Evangeline Walker Andrews and Charles McLean Andrews. New Haven: Yale UP, 1934.

Seward, Anna. *Original Sonnets on Various Subjects; and Odes Paraphrased from Horace.* 2nd ed. London, 1799.

———. *Memoirs of the Life of Dr. Darwin, Chiefly During his Residence at Lichfield, with Anecdotes of his Friends, and Criticisms on his Writings.* London: J. Johnson, 1804.

———. *The Poetical Works of Anna Seward; with Extracts from her Literary Correspondence.* 3 vols. Ed. Walter Scott. Edinburgh: Ballantyne and Co., 1810.

———. *Letters of Anna Seward: Written Between the Years 1784 and 1807.* 6 vols. Edinburgh: Constable and Co., 1811.

Shelley, Mary. *The Last Man.* 1826. Lincoln: U of Nebraska P, 2006.

———. *The Letters of Mary Wollstonecraft Shelley.* 3 vols. Ed. Betty T. Bennett. Baltimore: Johns Hopkins UP, 1980.

———. *The Journals of Mary Shelley, 1814–44.* Ed. Paula R. Feldman and Diana Scott-Kilvert. Baltimore: Johns Hopkins UP, 1987, 1995.

Shelley, Percy Bysshe. *Prometheus Unbound.* London: C. and J. Ollier, 1820.

———. *The Complete Works of Percy Bysshe Shelley.* 10 vols. Ed. Roger Ingpen and Walter E. Peck. New York: Charles Scribner's Sons, 1929.

———. *The Letters of Percy Bysshe Shelley.* 2 vols. Ed. Frederick L. Jones. Oxford: Clarendon P, 1964.

———. *Shelley's Prose.* Ed. David Lee Clark. London: Fourth Estate, 1988.

Smellie, William. *Philosophy of Natural History.* Edinburgh: printed for C. Elliot and T. Kay, T. Cadell, and G. G. J. and J. Robinsons, 1790.

Smith, Adam. *Essays on Philosophical Subjects.* Written circa 1758. London, 1795.

Smith, Charlotte. *Rural Walks; In Dialogues, intended for the Use of Young Persons.* 2 vols. London: T. Cadell and W. Davies, 1795.

———. *Conversations Introducing Poetry: Chiefly on Subjects of Natural History.* London: printed for J. Johnson, 1804.

———. *A Natural History of Birds, Intended Chiefly for Young Persons*. 2 vols. London: J. Sharpe, 1807, 1815.

———. *The Poems of Charlotte Smith*. Ed. Stuart Curran. New York: Oxford UP, 1993.

———. *The Collected Letters of Charlotte Smith*. Ed. Judith Phillips Stanton. Bloomington: Indiana UP, 2003.

Southey, Robert. *Omniana; or, Horae Otiosiores*. 2 vols. London, 1812.

Spenser, Edmund. *Complete Works of Edmund Spenser*. Ed. R. Morris. London: Macmillan, 1897.

Steele, Richard. *The Guardian* 12 (Mar. 25, 1713).

Stewart, Charles. *Elements of the Natural History of the Animal Kingdom*. 2 vols. 2nd ed. Edinburgh: printed for Bell & Bradfute; London: Longman, Hurst, Rees, Orme, and Browne, 1817.

Stockdale, Percival. *An Inquiry into the Nature and Genuine Laws of Poetry; Including a Particular Defence of the Writings, and Genius of Mr. Pope*. London: printed for N. Conant, 1778.

Stone, J. H., and Helen Maria Williams, *Copies of Original Letters Recently Written by Persons in Paris to Dr. Priestley in America*. 2nd ed. London, 1798.

Styles, John. *Lord Byron's Works Viewed in Connexion with Christianity, and the Obligation of Social Life*. London, 1824.

Tennyson, Hallam. *Alfred, Lord Tennyson; A Memoir by his Son*. 2 vols. Macmillan, 1897.

Thomson, James. *The Complete Poetical Works of James Thomson*. Ed. Henry Frowde. Oxford: Oxford UP, 1908.

Trapp, Joseph. "Of Didactic or Preceptive Poetry." 1711. *Lectures on Poetry*. London: printed for C. Hitch and C. Davis, 1742. 187–201.

Walker, John. *Elements of Geography*. 3rd ed. Dublin, 1797, 1788.

Warton, Joseph. *The Adventurer*, no. 63 (Tuesday, June 12, 1753).

Williams, Helen Maria. *A Tour in Switzerland; or, A View of the Present State of the Governments and Manners of those Cantons: With Comparative Sketches of the Present State of Paris*. 2 vols. London, 1798.

———. *Poems on Various Subjects: With Introductory Remarks on the Present State of Science and Literature in France*. London: G. and W. B. Whitaker, 1823.

———. *Letters from France*. 8 vols in 2. Facsimile reproduction with an introduction by Janet M. Todd. Delmar, NY: Scholars' Facsimiles and Reprints, 1975.

———. *A Tour in Switzerland*. Ed. Patrick Vincent and Florence Widmer-Schnyder. Paris: Diffusion France, 2011.

Wilton, Pleydell. *Geology and Other Poems*. London: printed for J. Hatchard, 1818.

Withering, William. "Introduction to the Study of Botany." *An Arrangement of British Plants, According to the Latest Improvements of the Linnaean System to Which is Prefixed an Easy Introduction to the Study of Botany. Illustrated By Copper Plates By William Withering*. 4 vols. 3rd ed. Birmingham: Robinson, 1796. 1–8.

Wollstonecraft, Mary. *A Critical Edition of Mary Wollstonecraft's A Vindication of the Rights of Woman: With Strictures on Political and Moral Subjects*. Ed. Ulrich H. Hardt. Troy, NY: Whitston Publishing Co., 1982.

Wordsworth, William. *The Prose Works of William Wordsworth.* Ed. W. J. B. Owen and Jane Worthington Smyser. 3 vols. Oxford: Clarendon P, 1974.

———. *The Prelude.* Ed. J. C. Maxwell. New Haven: Yale UP, 1981.

———. *William Wordsworth.* Ed. Stephen Gill. Oxford: Oxford UP, 1984.

Young, Edward. *Conjectures on Original Composition. In a Letter to the Author of Sir Charles Grandison.* London: printed for A. Millar, 1759.

SECONDARY SOURCES

Abrams, M. H. *The Mirror and the Lamp: Romantic Theory and the Critical Tradition.* 1953. London: Oxford UP, 1971.

———. "Structure and Style in the Greater Romantic Lyric." *From Sensibility to Romanticism: Essays Presented to Frederick A. Pottle.* Ed. Frederick W. Hilles and Harold Bloom. New York: Oxford UP, 1965. 527–60.

———. *Natural Supernaturalism: Tradition and Revolution in Romantic Literature.* New York: Norton, 1971.

Ahern, Stephen, ed. *Affect and Abolition in the Anglo-Atlantic, 1770–1830.* Burlington, VT: Ashgate, 2013.

Anderson, John M. "*Beachy Head:* The Romantic Fragment Poem as Mosaic." *Forging Connections: Women's Poetry from the Renaissance to Romanticism.* Ed. Anne K. Mellor, Felicity Nussbaum, and Jonathan F. S. Post. San Marino, CA: Huntington Library, 2002. 119–46.

Andrews, Corey E. "Scarred, Suffering Bodies: Eighteenth-Century Scottish Women Travellers on Slavery, Sentiment and Sensibility." *Women in Eighteenth-Century Scotland.* Ed. Katie Barclay and Deborah Simonton. Burlington, VT: Ashgate, 2013. 171–89.

Armitage, David. *The Ideological Origins of the British Empire.* New York: Cambridge UP, 2000.

Armstrong, Isobel. "Natural and National Monuments—Felicia Hemans's 'The Image in Lava': A Note." *Felicia Hemans: Reimagining Poetry in the Nineteenth Century.* Ed. Nanora Sweet and Julie Melnyk, 2001. 212–30.

Ashmun, Margaret. *The Singing Swan: An Account of Anna Seward and Her Acquaintance with Dr. Johnson, Boswell, and Others of Their Time* (New Haven: Yale UP, 1931.

Backscheider, Paula. *Eighteenth-Century Women Poets and Their Poetry: Inventing Agency, Inventing Genre.* Baltimore: Johns Hopkins UP, 2005.

Bailey, Edward Battersby. *James Hutton: The Founder of Modern Geology.* New York: Elsevier Publishing, 1967.

Barnard, Teresa. *Anna Seward: A Constructed Life: A Critical Biography.* Farnham: Ashgate, 2009.

———. "Anna Seward's Hidden Words: Female Interventions into Male Writing." *Women's Writing* 19.4 (Nov. 2012): 417–33.

———. "Anna Seward's 'Terrestrial Year': Women, Poetry, and Science in Eighteenth-Century England." *Partial Answers* 7.1 (Jan. 2009): 3–17.

Barrell, John. *The Political Theory of Painting from Reynolds to Hazlitt: "The Body of the Public."* New Haven: Yale UP, 1986.

Bate, Walter Jackson. *The Burden of the Past and the English Poet.* New York: Norton, 1970, 1972.

Battersby, Christine. *Gender and Genius: Toward a Feminist Aesthetic.* London: Women's P, 1989.

Beer, Gillian. *Darwin's Plots: Evolutionary Narrative in Darwin, George Eliot and Nineteenth-Century Fiction.* Cambridge: Cambridge UP, 1983, 2000.

Bellanca, Mary Ellen. *Daybooks of Discovery: Nature Diaries in Britain, 1770–1870.* Charlottesville: U of Virginia P, 2007.

Bending, Stephen. *Green Retreats: Women, Gardens, and Eighteenth-Century Culture.* Cambridge: Cambridge UP, 2013.

Benjamin, Marina. *Science and Sensibility: Gender and Scientific Enquiry, 1780–1945.* Oxford: Blackwell, 1991.

———, ed. *A Question of Identity: Women, Science, and Literature.* New Brunswick: Rutgers UP, 1993.

Bennett, Betty T. "Radical Imaginings: Mary Shelley's *The Last Man.*" *Wordsworth Circle* 26.3 (Summer 1995): 147–52.

Berns, Ute. *Science, Politics, and Friendship in the Works of Thomas Lovell Beddoes.* Newark: U of Delaware P, 2012.

Bewell, Alan. *Wordsworth and the Enlightenment: Nature, Man, and Society in the Experimental Poetry.* New Haven: Yale UP, 1989.

———. "'Jacobin Plants': Botany as Social Theory in the 1790s." *Wordsworth Circle* 20.3 (Summer 1989): 132–39.

———. "'On the Banks of the South Sea': Botany and Sexual Controversy in the Late Eighteenth Century." *Visions of Empire: Voyages, Botany, and Representations of Nature.* Ed. David Philip Miller and Peter Hanns Reill. Cambridge: Cambridge UP, 1996. 173–93.

———. *Romanticism and Colonial Disease.* Baltimore: John Hopkins UP, 1999.

———. "Jefferson's Thermometer: Colonial Biogeographical Constructions of the Climate of America." *Romantic Science: The Literary Forms of Natural History.* Ed. Noah Heringman. Albany: SUNY UP, 2003. 111–38.

———. "Erasmus Darwin's Cosmopolitan Nature." *ELH* 76.1 (Spring 2009): 19–48.

Bohls, Elizabeth A. *Women Travel Writers and the Language of Aesthetics, 1716–1818.* Cambridge: Cambridge UP, 1995.

Bowler, Peter J. *Evolution: The History of an Idea.* Berkeley: U of California P, 1984.

Boyle, Deborah. "Astell and Cartesian 'Scientia.'" *The New Science and Women's Literary Discourse: Prefiguring Frankenstein.* Ed. Judy A. Hayden. New York: Palgrave, 2011. 99–112.

Brown, Vincent. *The Reaper's Garden: Death and Power in the World of Atlantic Slavery.* Cambridge: Harvard UP, 2008.

Browne, Janet. "Botany for Gentlemen: Erasmus Darwin and 'The Loves of the Plants.'" *Isis* 80.4 (1989): 593–621.

Buckland, Adelene. *Novel Science: Fiction and the Invention of Nineteenth-Century Geology.* Chicago: U of Chicago P, 2013.

Burwick, Frederick. "Romantic Theories of Translation." *The Wordsworth Circle* 39.3 (Summer 2008): 68–75.

Butler, Marilyn. "Frankenstein and Radical Science." *Frankenstein: Norton Critical Edition.* Ed. J. Paul Hunter. New York: Norton, 1996. 302–13.

Cameron, Lauren. "Mary Shelley's Malthusian Objections in *The Last Man*." *Nineteenth-Century Literature* 67.2 (Sept. 2012): 177–203.

Cantor, Paul A. "The Apocalypse of Empire: Mary Shelley's *The Last Man*." *Iconoclastic Departures: Mary Shelley after Frankenstein*. Ed. Syndy M. Conger, Frederick S. Frank, and Gregory O'Dea. Madison, NJ: Fairleigh Dickinson UP, 1997. 193–211.

Casid, Jill H. *Sowing Empire: Landscape and Colonization*. Minneapolis: U of Minnesota P, 2005.

Chalker, John. *The English Georgic: A Study in the Development of Form*. London: Routledge and Kegan Paul, 1969.

Chapple, J. A. V. *Science and Literature in the Nineteenth Century*. London: Macmillan, 1986.

Clarke, Norma. "Anna Seward: Swan, Duckling or Goose?" *British Women's Writing in the Long Eighteenth Century*. Ed. Jennie Batchelor and Cora Kaplan. New York: Palgrave, 2005. 34–47.

Clymer, Lorna. Introduction to *Ritual, Routine, and Regime: Repetition in Early Modern British and European Cultures*. Ed. Lorna Clymer. Toronto: U of Toronto P, 2006. 3–20.

Cody, Lisa Foreman. *Birthing the Nation: Sex, Science, and the Conception of Eighteenth-Century Britons*. Oxford: Oxford UP, 2005.

Cohen, I. Bernard. *Revolution in Science*. Cambridge: Harvard UP, 1985.

Colley, Linda. *Britons: Forging the Nation, 1707–1837*. 2nd ed. New Haven: Yale UP, 1992.

Cook, Elizabeth Heckendorn. "Charlotte Smith and 'The Swallow': Migration and Romantic Authorship." *Huntington Library Quarterly* 72.1 (2009): 48–67.

Cosslett, Tess. *The "Scientific Movement" and Victorian Literature*. Sussex: Harvester P, 1982.

Cousins, A. D. "Marvell's Devout Mythology of the New World: Homeland and Home in 'Bermudas'." *Parergon* 30.1 (2013): 203–19.

Crawford, Rachel. *Poetry, Enclosure, and the Vernacular Landscape, 1700–1830*. Cambridge: Cambridge UP, 2002.

Curran, Stuart. *Poetic Form and British Romanticism*. New York: Oxford UP, 1986.

———. "The I Altered." *Romanticism and Feminism*. Ed. Anne K. Mellor. Bloomington: Indiana UP, 1988. 185–207.

Dale, Peter Allan. *In Pursuit of a Scientific Culture: Science, Art, and Society in the Victorian Age*. Madison: U of Wisconsin P, 1989.

Damián, Jessica. "Helen Maria Williams's Personal Narrative of Travels from *Peru* (1784) to *Peruvian Tales* (1823)." *Nineteenth Century Gender Studies* 3.2 (Summer 2007): http://www.ncgsjournal.com/issue32/damian.htm.

Dawson, Gowan. *Darwin, Literature and Victorian Respectability*. Cambridge: Cambridge UP, 2007.

Dayfoot, Arthur Charles. *The Shaping of the West Indian Church, 1492–1962*. Gainesville: UP of Florida, 1999.

De Carolis, Ernesto, and Giovanni Patricelli. *Vesuvius A.D. 79: The Destruction of Pompeii and Herculaneum*. Los Angeles: Getty P, 2003.

DeLucia, JoEllen. "Local Poetry in the Midlands: Francis Mundy's *Needwood*

Forest and Anna Seward's Lichfield Poems." *Representing Place in British Literature and Culture, 1660-1830: From Local to Global*. Ed. Evan Gottlieb and Juliet Shields. Burlington, VT: Ashgate, 2013. 155-71.

Dettelbach, Michael. "Global Physics and Aesthetic Empire: Humboldt's Physical Portrait of the Tropics." *Visions of Empire: Voyages, Botany, and Representations of Nature*. Ed. David Philip Miller and Peter Hanns Reill. New York: Cambridge UP, 1996. 258-92.

———. "Humboldtian Science." *Cultures of Natural History*. Ed. N. Jardine, J. A. Secord, E. C. Spary. Cambridge: Cambridge UP, 1996, 2000. 287-304.

Dolan, Elizabeth A. *Seeing Suffering in Women's Literature of the Romantic Era*. Burlington, VT: Ashgate, 2008.

Doody, Margaret Anne. "Sensuousness in the Poetry of Eighteenth-Century Women Poets." *Women's Poetry in the Enlightenment: The Making of a Canon, 1730-1820*. Ed. Isobel Armstrong and Virginia Blain. New York: St. Martin's P, 1999. 3-32.

Drouin, Jean-Marc, and Bernadette Bensaude-Vincent. "Nature for the People." *Cultures of Natural History*. Ed. N. Jardine, J. A. Secord, and E. C. Spary. Cambridge: Cambridge UP, 1996. 408-25.

Dwyer, Eugene. *Pompeii's Living Statues: Ancient Roman Lives Stolen from Death*. Ann Arbor: U of Michigan P, 2010.

Dyce, Alexander, ed. *Recollections of the Table-Talk of Samuel Rogers*. New Southgate, 1887.

Edwards, Pamela. *The Statesman's Science: History, Nature, and Law in the Political Thought of Samuel Taylor Coleridge*. New York: Columbia UP, 2004.

Fara, Patricia. *Erasmus Darwin: Sex, Science, and Serendipity*. Oxford: Oxford UP, 2012.

Feingold, Richard. *Nature and Society: Later Eighteenth-Century Uses of the Pastoral and Georgic*. New Brunswick, NJ: Rutgers UP, 1978.

Fletcher, Loraine. *Charlotte Smith: A Critical Biography*. New York: St. Martin's, 1998.

Frohock, Richard. *Heroes of Empire: The British Imperial Protagonist in America, 1596-1764*. Newark: U of Delaware P, 2004.

Fulford, Tim, Debbie Lee, and Peter Kitson. *Literature, Science, and Exploration in the Romantic Era: Bodies of Knowledge*. Cambridge: Cambridge UP, 2004.

Gallagher, William B. *When Dinosaurs Roamed New Jersey*. New Brunswick: Rutgers UP, 1997.

Gates, Barbara T. *Kindred Nature: Victorian and Edwardian Women Embrace the Living World*. Chicago: U of Chicago P, 1998.

George, Sam. *Botany, Sexuality, and Women's Writing, 1760-1830: From Modest Shoot to Forward Plant*. Manchester: Manchester UP, 2007.

Gilbert, Sandra, and Susan Gubar. *The Madwoman in the Attic: The Woman Writer and the Nineteenth-Century Literary Imagination*. New Haven: Yale UP, 1979.

Gilmore, John. *The Poetics of Empire: A Study of James Grainger's The Sugar Cane (1764)*. London: Athlone P, 2000.

Girten, Kristin M. "Unsexed Souls: Natural Philosophy as Transformation in Eliza Haywood's *Female Spectator*." *Eighteenth-Century Studies* 43.1 (Fall 2009): 55-74.

Gladstone, Hugh S. *Maria Riddell, The Friend of Burns.* N.p.: Dumfries, 1915.

Goellnicht, Donald C. *The Poet-Physician: Keats and Medical Science.* Pittsburgh: U of Pittsburgh P, 1984.

Goodman, Kevis. *Georgic Modernity and British Romanticism: Poetry and the Mediation of History.* Cambridge: Cambridge UP, 2004.

Gordon, Elizabeth Oke. *The Life and Correspondence of William Buckland.* London: John Murray, 1894.

Gottlieb, Evan. *Feeling British: Sympathy and National Identity in Scottish and English Writing, 1707–1832.* Lewisburg: Bucknell UP, 2007.

———. *Romantic Globalism: British Literature and Modern World Order, 1750–1830.* Columbus: Ohio State UP, 2014.

Gould, Stephen Jay. *Hen's Teeth and Horse's Toes.* New York: Norton, 1983.

Grant, Peter R. Introduction to *Evolution on Islands.* New York: Oxford UP, 1998. 1–17.

Gravil, Richard. "Helen Maria Williams: Wordsworth's Revolutionary Anima." *Wordsworth Circle* 40.1 (2009): 55–64.

Gregory, Desmond. *The Beneficent Usurpers: A History of the British in Madeira.* Toronto: Associated U Presses, 1988.

Grove, Richard H. *Green Imperialism: Colonial Expansion, Tropical Island Edens, and the Origins of Environmentalism, 1600–1860.* New York: Cambridge UP, 1995.

Guest, Harriet. *Small Change: Women, Learning, Patriotism, 1750–1810.* Chicago: U of Chicago P, 2000.

Hartman, Geoffrey. *Wordsworth's Poetry, 1787–1814.* New Haven: Yale UP, 1964.

Hayden, Judy A. "Introduction: Women, Education, and the Margins of Science." *The New Science and Women's Literary Discourse: Prefiguring Frankenstein.* Ed. Judy A. Hayden. New York: Palgrave, 2011. 1–15.

Heringman, Noah. "The Rock Record and Romantic Narratives of the Earth." *Romantic Science: The Literary Forms of Natural History.* Ed. Noah Heringman. Albany: SUNY UP, 2003. 53–84.

———. *Romantic Rocks, Aesthetic Geology.* Ithaca: Cornell UP, 2004.

———. "Natural History in the Romantic Period." *A Concise Companion to the Romantic Age.* Ed. Jon Klancher.Chichester: Wiley-Blackwell, 2009. 141–67.

———. *Sciences of Antiquity: Romantic Antiquarianism, Natural History, and Knowledge Work.* Corby: Oxford UP, 2013.

Hollocher, Hope. "Island Hopping in *Drosophilia*: Genetic Patterns vs. Evolutionary Processes." *Evolution on Islands.* Ed. Peter R. Grant. New York: Oxford UP, 1998. 124–41.

Holmes, Richard. *The Age of Wonder: How the Romantic Generation Discovered the Beauty and Terror of Science.* New York: Random House, 2008.

Italia, Iona. *The Rise of Literary Journalism in the Eighteenth Century: Anxious Employment.* London: Routledge, 2005.

Jeffrey, Lloyd N. "Cuvierian Catastrophism in Shelley's *Prometheus Unbound* and 'Mont Blanc.'" *South Central Bulletin,* Winter 1978, 148–52.

Jones, Christine Kenyon. "'When This World Shall Be Former': Catastrophism as Imaginative Theory for the Young Romantics." *Érudit: Romanticism on the Net* 24 (Nov. 2001): http://id.erudit.org/iderudit/006000ar.

Kairoff, Claudia Thomas. *Anna Seward and the End of the Eighteenth Century.* Baltimore: Johns Hopkins UP, 2012.

Karkoulis, Dimitri. "'They Pluck'd the Tree of Science/And Sin': Byron's Cain and the Science of Sacrilege." *European Romantic Review* 18.2 (Apr. 2007): 273–81.

Kaul, Suvir. *Poems of Nation, Anthems of Empire: English Verse in the Long Eighteenth Century.* Charlottesville: UP of Virginia, 2000.

Keith, Jennifer. *Poetry and the Feminine from Behn to Cowper.* Newark: U of Delaware P, 2005.

Kelley, Theresa M. "Romantic Exemplarity: Botany and 'Material' Culture." *Romantic Science: The Literary Forms of Natural History.* Ed. Noah Heringman. Albany: SUNY P, 2003. 223–54.

———. *Clandestine Marriage: Botany and Romantic Culture.* Baltimore: Johns Hopkins UP, 2012.

Kelly, Gary. "Death and the Matron: Felicia Hemans, Romantic Death, and the Founding of the Modern Liberal State." *Felicia Hemans: Reimagining Poetry in the Nineteenth Century.* Ed. Nanora Sweet and Julie Melnyk, 2001. 196–211.

———, ed. *Felicia Hemans: Selected Poems, Prose, and Letters.* Ontario: Broadview P, 2002.

Kennedy, Deborah. *Helen Maria Williams and the Age of Revolution.* Lewisburg: Bucknell UP, 2002.

King-Hele, Desmond. *Erasmus Darwin: A Life of Unequalled Achievement.* London: Giles de la Mare Publishers Ltd., 1999.

Kirkley, Laura. "Translating Rousseauism: Transformations of Bernardin de Saint-Pierre's *Paul et Virginie* in the Works of Helen Maria Williams and Maria Edgeworth." *Readers, Writers, Salonnières: Female Networks in Europe, 1700–1900.* Oxford: Peter Lang, 2011. 93–118.

Klancher, Jon. *Transfiguring the Arts and Sciences: Knowledge and Cultural Institutions in the Romantic Age.* Cambridge: Cambridge UP, 2013.

Knight, David. *Ordering the World: A History of Classifying Man.* London: Burnett Books, 1981.

Koerner, Lisbet. *Linnaeus: Nature and Nation.* Cambridge, Mass.: Harvard UP, 1999.

Kucich, John. "Scientific Ascendancy." *A Companion to the Victorian Novel.* Ed. Patrick Brantlinger and William B. Thesing. Malden, MA: Blackwell, 2005. 119–36.

Kunzle, David. "World Upside Down: The Iconography of a European Broadsheet Type." *The Reversible World: Symbolic Inversion in Art and Society.* Ed. Barbara A. Babcock. Ithaca: Cornell UP, 1978. 39–94.

Labbe, Jacqueline. "'Transplanted into a more congenial soil': Footnoting the Self in the Poetry of Charlotte Smith." *Ma(r)king the Text: The Presentation of Meaning on the Literary Page.* Ed. Bray, Handley, and Henry. Burlington, VT: Ashgate, 2000. 71–86.

———. *Charlotte Smith: Romanticism, Poetry, and the Culture of Gender.* New York: Manchester UP, 2003.

———. "'The absurdity of animals having the passions and the faculties of man':

Charlotte Smith's Fables (1807)." *European Romantic Review* 19.2 (Apr. 2008): 157–62.

———. "The Hybrid Poems of Smith and Wordsworth: Questions and Disputes." *European Romantic Review* 20.2 (Apr. 2009): 219–26.

Landry, Donna. "Green Languages?: Women Poets as Naturalists in 1653 and 1807." *Forging Connections: Women's Poetry from the Renaissance to Romanticism.* Ed. Anne K. Mellor, Felicity Nussbaum, and Jonathan F. S. Post. San Marino, CA: Huntington Library, 2002. 39–61.

Lawlor, Clark. "Poetry and Science." *A Companion to Eighteenth-Century Poetry.* Ed. Christine Gerrard. Malden, MA: Blackwell P, 2006. 38–52.

Leader, Zachary. *Revision and Romantic Authorship.* Oxford: Clarendon P, 1996.

Leask, Nigel. "Mont Blanc's Mysterious Voice: Shelley and Huttonian Earth Science." *The Third Culture: Literature and Science.* Ed. Elinor S. Shaffer. New York: Walter de Gruyter, 1998. 182–203.

———. "Salons, Alps, and Cordilleras: Helen Maria Williams, Alexander von Humboldt, and the Discourse of Romantic Travel." *Women, Writing, and the Public Sphere, 1700–1830.* Ed. Elizabeth Eger, Charlotte Grant, Clíona Ó Gallchoir, and Penny Warburton. Cambridge: Cambridge UP, 2001. 217–35.

———. *Curiosity and Aesthetics of Travel Writing, 1770–1840.* Oxford: Oxford UP, 2002.

Leighton, Angela. *Victorian Women Poets: Writing against the Heart.* Charlottesville: UP of Virginia, 1992.

Levere, Trevor H. *Poetry Realized in Nature: Samuel Taylor Coleridge and Early Nineteenth-Century Science.* Cambridge: Cambridge UP, 1981.

Levine, George. *Darwin and the Novelists: Patterns of Science in Victorian Fiction.* Cambridge: Harvard UP, 1988.

Lew, Joseph W. "The Plague of Imperial Desire: Montesquieu, Gibbon, Brougham, and Mary Shelley's *The Last Man.*" *Romanticism and Colonialism: Writing and Empire, 1780–1830.* Ed. Tim Fulford and Peter J. Kitson. Cambridge: Cambridge UP, 1998. 261–78.

Lokke, Kari E. "*The Last Man.*" *The Cambridge Companion to Mary Shelley.* Ed. Esther Schor. Cambridge: Cambridge UP, 2003. 116–34.

Looser, Devoney. "Why I'm Still Writing Women's Literary History." *The Minnesota Review* 71–72 (Winter/Spring 2009): 220–27.

Losos, Jonathan B. "Ecological and Evolutionary Determinants of the Species-Area Relationship in Caribbean Anoline Lizards." *Evolution on Islands.* Ed. Peter R. Grant. New York: Oxford UP, 1998. 210–24.

Low, Anthony. *The Georgic Revolution.* Princeton: Princeton UP, 1985.

Lundeen, Kathleen. "'When Life Becomes Art'—On Hemans's 'Image in Lava'." *Érudit: Romanticism on the Net* 29–30 (Feb.–May 2003): http://id.erudit.org /iderudit/007716ar.

Macfarlane, Robert. *Original Copy: Plagiarism and Originality in Nineteenth-Century Literature.* Oxford: Oxford UP, 2007.

Mackay, David. "Agents of Empire: The Banksian Collectors and Evaluation of New Land." *Visions of Empire: Voyages, Botany, and Representations of Nature.* Ed. David Philip Miller and Peter Hanns Reill. Cambridge: Cambridge UP, 1996. 38–54.

Macnaghten, Angus. *Burns' Mrs. Riddell: A Biography.* Peterhead: Volurna P, 1975.

Mahood, M. M. *The Poet as Botanist.* New York: Cambridge UP, 2008.

Mandell, Laura. *Misogynous Economies: The Business of Literature in Eighteenth-Century Britain.* Lexington: UP of Kentucky, 1999.

Martin, Alison E. "'These Changes and Accessions of Knowledge': Translation, Scientific Travel Writing and Modernity—Alexander von Humboldt's *Personal Narrative.*" *Studies in Travel Writing* 15.1 (Feb. 2011): 39–51.

Mazzeo, Tilar J. *Plagiarism and Literary Property in the Romantic Period.* Philadelphia: U of Pennsylvania P, 2007.

McCarthy, William. "'We Hoped the Woman was Going to Appear': Repression, Desire, and Gender in Anna Letitia Barbauld's Early Poems." *Romantic Women Writers: Voices and Countervoices.* Ed. Paula R. Feldman and Theresa M. Kelley. Hanover: UP of New England, 1995. 113–37.

———. *Anna Letitia Barbauld: Voice of Enlightenment.* Baltimore: Johns Hopkins UP, 2008.

McIntyre, Ian. *Dirt and Deity: A Life of Robert Burns.* London: Harper Collins, 1995.

McPhee, John. *Basin and Range.* New York: Farrar, Straus & Giroux. 1981.

Mellor, Anne K. "Frankenstein: A Feminist Critique of Science." *One Culture: Essays in Science and Literature.* Ed. George Levine and Alan Rauch. Madison: U of Wisconsin P, 1987. 287–312.

———. "Possessing Nature: The Female in *Frankenstein.*" *Romanticism and Feminism.* Ed. Anne K. Mellor. Bloomington: Indiana UP, 1988. 220–32.

———. *Mary Shelley: Her Life, Her Fiction, Her Monsters.* New York: Routledge, 1988, 1989.

———. "A Feminist Critique of Science." *Critical Essays on Mary Wollstonecraft Shelley.* Ed. Mary Lowe-Evans. New York: G. K. Hall, 1998. 62–87

———. *Mothers of the Nation: Women's Political Writing in England, 1780–1830.* Bloomington: Indiana UP, 2000.

———. "The Baffling Swallow: Gilbert White, Charlotte Smith and the Limits of Natural History." *Nineteenth-Century Contexts* 31.4 (Dec. 2009): 299–309.

Melnyk, Julie. "William Wordsworth and Felicia Hemans." *Fellow Romantics: Male and Female British Writers, 1790–1835.* Ed. Beth Lau, 2009. 139–58

Menke, Richard. "The Political Economy of Fruit: Goblin Market." *The Culture of Christina Rossetti: Female Poetics and Victorian Contexts.* Ed. Mary Arseneau, Antony Harrison, and Lorraine Kooistra. Athens: Ohio UP, 1999. 105–36.

Miller, Mary Ashburn. "Mountain, Become a Volcano: The Image of the Volcano in the Rhetoric of the French Revolution." *French Historical Studies* 32.4 (Fall 2009): 555–85.

Moore, Lisa L. *Sister Arts: The Erotics of Lesbian Landscapes.* Minneapolis: U of Minnesota P, 2011.

Morris, Michael. *Scotland and the Caribbean, c. 1740–1833: Atlantic Archipelagos.* New York: Routledge, 2015.

Murphy, Patricia. *In Science's Shadow: Literary Constructions of Late Victorian Women.* Columbia: U of Missouri P, 2006.

Nicolson, Malcolm. Introduction to Alexander von Humboldt, *Personal Narrative of a Journey to the Equinoctial Region of the New Continent.* London: Penguin, 1995.

O'Connor, Ralph. "Mammoths and Maggots: Byron and the Geology of Cuvier." *Romanticism* 5.1 (1999): 26–42.

———. "Kirkdale Cave and the Poetry of William Buckland." *Studies in Speleology* 14 (2006): 39–41.

———. *The Earth on Show: Fossils and the Poetics of Popular Science, 1802–1856.* Chicago: U of Chicago P, 2007.

———. "Byron's Afterlife and the Emancipation of Geology." *Liberty and Poetic Licence: New Essays on Byron.* Ed. Bernard Beatty, Tony Howe, and Charles E. Robinson. Liverpool: Liverpool UP, 2008. 147–64.

Oldroyd, David R. *Thinking about the Earth: A History of Ideas in Geology.* Cambridge: Harvard UP, 1996.

O'Shaughnessy, Andrew Jackson. *An Empire Divided: The American Revolution and the British Caribbean.* Philadelphia: U of Pennsylvania P, 2000.

Page, Judith W., and Elise L. Smith. *Women, Literature, and the Domesticated Landscape: England's Disciples of Flora, 1780–1870.* Cambridge: Cambridge UP, 2011.

Parrish, Susan Scott. *American Curiosity: Cultures of Natural History in the Colonial British Atlantic World.* Chapel Hill: U of North Carolina P, 2006.

Pascoe, Judith. "Female Botanists and the Poetry of Charlotte Smith." *Re-Visioning Romanticism: British Women Writers, 1776–1837.* Ed. Carol Shiner Wilson and Joel Haefner. Philadelphia: U of Pennsylvania P, 1994. 193–209.

———. "'Unsex'd Females': Barbauld, Robinson, and Smith." *The Cambridge Companion to English Literature, 1740–1830.* Ed. Thomas Keymer and Jon Mee. Cambridge: Cambridge UP, 2004. 211–26.

———. *The Hummingbird Cabinet: A Rare and Curious History of Romantic Collectors.* Ithaca: Cornell UP, 2006.

Pearson, Jacqueline. *Women's Reading in Britain, 1750–1835: A Dangerous Recreation.* Cambridge: Cambridge UP, 1999.

Pérez-Mejía, Ángela. *A Geography of Hard Times: Narratives about Travel to South America, 1780–1849.* Trans. Dick Cluster. Albany: SUNY P, 2004.

Peterfreund, Stuart. *William Blake in a Newtonian World: Essays on Literature as Art and Science.* Norman: U Oklahoma P, 1998.

Peterson, Linda H. *Becoming a Woman of Letters: Myths of Authorship and Facts of the Victorian Market.* Princeton: Princeton UP, 2009.

Pettitt, Clare. *Patent Inventions: Intellectual Property and the Victorian Novel.* Oxford: Oxford UP, 2004.

Pinch, Adela. *Strange Fits of Passion: Epistemologies of Emotion, Hume to Austen.* Stanford: Stanford UP, 1996.

Plank, Jeffrey. "John Aikin on Science and Poetry." *Studies in Burke and His Time* 18.1 (1977): 167–78.

Porter, Dahlia. "From Nosegay to Specimen Cabinet: Charlotte Smith and the Labour of Collecting." *Charlotte Smith in British Romanticism.* Ed. Jacqueline Labbe. London: Pickering and Chatto, 2008. 29–44.

———. "Formal Relocations: The Method of Southey's *Thalaba the Destroyer* (1801)." *European Romantic Review* 20.5 (Dec. 2009): 671–79.

Pratt, Mary Louise. *Imperial Eyes: Travel Writing and Transculturation.* New York: Routledge, 1992.

Robinson, Daniel. "Reviving the Sonnet: Women Romantic Poets and the Sonnet Claim." *European Romantic Review* 6.1 (Summer 1995): 98–127.

Roe, Nicholas, ed. *Samuel Taylor Coleridge and the Sciences of Life.* Oxford: Oxford UP, 2001.

Roger, Jacques. *Buffon: A Life in Natural History.* Trans. Sarah Lucille Bonnefoi. Ithaca: Cornell UP, 1997.

Ross, Catherine E. "'Twin Labourers and Heirs of the Same Hopes': The Professional Rivalry of Humphry Davy and William Wordsworth." *Romantic Science.* Ed. Noah Heringman. Albany: SUNY P, 2003. 23–52.

Ross, Trevor. "Two Ways of Looking at a Canon." *Eighteenth-Century Life* 21.3 (Nov. 1997): 90–93.

———. *The Making of the English Literary Canon from the Middle Ages to the Late Eighteenth Century.* Montreal: McGill-Queen's UP, 1998.

Rudwick, Martin J. *Scenes from Deep Time: Early Pictorial Representations of the Prehistoric World.* Chicago: U of Chicago P, 1992.

———. *Georges Cuvier, Fossil Bones, and Geological Catastrophes: New Translations and Interpretations of the Primary Texts.* Chicago: U of Chicago P, 1997.

Rupke, Nicolaas A. *The Great Chain of History: William Buckland and the English School of Geology, 1814–1849.* Oxford: Clarendon P, 1983.

Russell, Gillian, and Clara Tuite, eds. *Romantic Sociability: Social Networks and Literary Culture in Britain, 1770–1840.* Cambridge: Cambridge UP, 2002.

Ruston, Sharon. *Shelley and Vitality.* Houndmills: Palgrave, 2005.

———. *Creating Romanticism: Case Studies in the Literature, Science, and Medicine of the 1790s.* Houndmills: Palgrave Macmillan, 2013.

Ruwe, Donelle R. "Charlotte Smith's Sublime: Feminine Poetics, Botany and *Beachy Head.*" *Prism(s): Essays in Romanticism* 7 (1999): 117–32.

Sachs, Aaron. *The Humboldt Current: Nineteenth-Century Exploration and the Roots of American Environmentalism.* New York: Viking, 2006.

Sambrook, James. *James Thomson, 1700–1748: A Life.* Oxford: Clarendon P, 1991.

Scheinberg, Cynthia. *Women's Poetry and Religion in Victorian England: Jewish Identity and Christian Culture.* Cambridge: Cambridge UP, 2002.

Schiebinger, Londa. *The Mind Has No Sex? Women in the Origins of Modern Science.* Cambridge: Harvard UP, 1989.

———. *Nature's Body: Gender in the Making of Modern Science.* Boston: Beacon P, 1993.

———. "Gender and Natural History." *Cultures of Natural History.* Ed. N. Jardine, J. A. Secord, and E. C. Spary. Cambridge: Cambridge UP, 1996. 163–77.

———. *Plants and Empire: Colonial Bioprospecting in the Atlantic World.* Cambridge: Harvard UP, 2004.

Sheffield, Suzanne Le-May. *Revealing New Worlds: Three Victorian Women Naturalists.* London: Routledge, 2001.

Shteir, Ann. *Cultivating Women, Cultivating Science: Flora's Daughters and Botany in England, 1760–1860.* Baltimore: Johns Hopkins UP, 1996.

———. "Green-Stocking or Blue? Science in Three Women's Magazines, 1800–50." *Culture and Science in the Nineteenth-Century Media.* Ed. Louise Henson, Geoffrey Cantor, Gowan Dawson, Richard Noakes, Sally Shuttleworth, and Jonathan R. Topham. Aldershot: Ashgate, 2004. 3–14.

Shuttleworth, Sally. *George Eliot and Nineteenth-Century Science: The Make-Believe of a Beginning.* Cambridge: Cambridge UP, 1984.

Siskin, Clifford. *The Historicity of Romantic Discourse.* New York: Oxford UP, 1988.

Sloan, Phillip. "The Gaze of Natural History." *Inventing Human Science: Eighteenth-Century Domains.* Ed. Christopher Fox, Roy Porter, and Robert Wokler. Berkeley: U of California P, 1995. 112–51.

Sommer, Marianne. *Bones and Ochre: The Curious Afterlife of the Red Lady of Paviland.* Cambridge: Harvard UP, 2007.

Stafford, Fiona J. *The Last of the Race: The Growth of a Myth from Milton to Darwin.* Oxford: Clarendon P, 1994.

Stillinger, Jack. *Multiple Authorship and the Myth of Solitary Genius.* New York: Oxford UP, 1991.

Sussman, Charlotte. "'Islanded in the World': Cultural Memory and Human Mobility in *The Last Man.*" *PMLA* 118.2 (2003): 286–301.

Tayebi, Kandi. "Undermining the Eighteenth-Century Pastoral: Rewriting the Poet's Relationship to Nature in Charlotte Smith's Poetry." *European Romantic Review* 15.1 (Mar. 2004): 131–50.

Terry, Richard. *The Plagiarism Allegation in English Literature from Butler to Sterne.* Basingstoke: Palgrave, 2010.

Thomas, Sophie. "The Ends of the Fragment, the Problem of the Preface: Proliferation and Finality in *The Last Man.*" *Mary Shelley's Fictions: From Frankenstein to Falkner.* Ed. Michael Eberle-Sinatra. Basingstoke: Macmillan, 2000. 22–38.

Thomas, W. K., and Warren U. Ober, *A Mind For Ever Voyaging: Wordsworth at Work, Portraying Newton and Science.* Edmonton: U of Alberta P, 1989.

Tobin, Beth Fowkes. *Colonizing Nature: The Tropics in British Arts and Letters, 1760–1820.* Philadelphia: U of Pennsylvania P, 2005.

———. *The Duchess's Shells: Natural History Collecting in the Age of Cook's Voyages.* New Haven: Yale UP, 2014.

Todd, Dennis. *Imagining Monsters: Miscreations of the Self in Eighteenth-Century England.* Chicago: U of Chicago P, 1995.

Turner, Frank. *Contesting Cultural Authority: Essays in Victorian Intellectual Life.* Cambridge: Cambridge UP, 1993.

———. "The Late Victorian Conflict of Science and Religion as an Event in Nineteenth-Century Intellectual and Cultural History." *Science and Religion: New Historical Perspectives.* Ed. Thomas Dixon, Geoffrey Cantor, and Stephen Pumfrey. Cambridge: Cambridge UP, 2010. 87–110.

Turner, Katherine. *British Travel Writers in Europe 1750–1800: Authorship, Gender and National Identity.* Aldershot: Ashgate, 2001.

Uglow, Jenny. *The Lunar Men: The Friends Who Made the Future, 1730–1810.* New York: Faber and Faber, 2002.

———. "But What about the Women? The Lunar Society's Attitude to Women and Science, and to the Education of Girls." *The Genius of Erasmus Darwin.* Ed. C. U. M. Smith and Robert Arnott. Aldershot: Ashgate, 2005. 163–78.

Vasbinder, Samuel Holmes. *Scientific Attitudes in Mary Shelley's Frankenstein.* Ann Arbor, MI: UMI Research P, 1984.

Wallace, Anne D. "Picturesque Fossils, Sublime Geology?: The Crisis of Authority in Charlotte Smith's Beachy Head." *European Romantic Review* 13 (2002): 77–93.

Wallace, Tara Ghoshal. *Imperial Characters: Home and Periphery in Eighteenth-Century Literature.* Lewisburg: Bucknell UP, 2010.

Walls, Laura Dassow. *The Passage to Cosmos: Alexander von Humboldt and the Shaping of America.* Chicago: U of Chicago P, 2009.

Webb, Samantha. "Reading the End of the World: *The Last Man,* History, and the Agency of Romantic Authorship." *Mary Shelley in Her Times.* Ed. Betty T. Bennett and Stuart Curran. Baltimore: Johns Hopkins UP, 2000. 119–33.

Wheeler, Roxann. *The Complexion of Race: Categories of Difference in Eighteenth-Century British Culture.* Philadelphia: U of Pennsylvania P, 2000.

Williams, Glyn. *The Death of Captain Cook: A Hero Made and Unmade.* Cambridge: Harvard UP, 2008.

Wilson, Kathleen. *The Island Race: Englishness, Empire and Gender in the Eighteenth Century.* New York: Routledge, 2003.

Wolfson, Susan. "Charlotte Smith's *Emigrants:* Forging Connections at the Borders of a Female Tradition." *Forging Connections: Women's Poetry from the Renaissance to Romanticism.* Ed. Anne K. Mellor, Felicity Nussbaum, and Jonathan F. S. Post. San Marino, CA: Huntington Library, 2002. 81–118.

———. *Borderlines: The Shiftings of Gender in British Romanticism.* Stanford: Stanford UP, 2006.

———, ed. *Felicia Hemans: Selected Poems, Letters, Reception Materials.* Princeton: Princeton UP, 2000.

Wood, Gillen D'Arcy. "The Female Penseroso: Anna Seward, Sociable Poetry, and the Handelian Consensus." *Modern Language Quarterly* 67.4 (Dec. 2006): 451–77.

Wyatt, John. *Wordsworth and the Geologists.* Cambridge: Cambridge UP, 1995.

Young, Robert J. C. *Colonial Desire: Hybridity in Theory, Culture and Race.* London: Routledge, 1995.

INDEX

✛